# Living with Wood

DR. SERI C. ROBINSON
with photomicrographs by
DR. SARATH M. VEGA GUTIERREZ

# Living with Wood

DR. SERI C. ROBINSON
with photomicrographs by
DR. SARATH M. VEGA GUTIERREZ

A GUIDE FOR TOYMAKERS, HOBBYISTS, CRAFTERS, AND PARENTS

4880 Lower Valley Road • Atglen, PA 19310

Library of Congress Control Number: 2019947416



Edited by Ian Robertson
Designed by Ashley Millhouse
Cover design by Ro Shillingford
Type set in Caecilia LT Std/Whitney
ISBN: 978-0-7643-5935-4
Printed in China

Published by Schiffer Publishing, Ltd.
4880 Lower Valley Road
Atglen, PA 19310
Phone: (610) 593-1777; Fax: (610) 593-2002
E-mail: Info@schifferbooks.com
Web: www.schifferbooks.com

# Acknowledgments

The amazing microscopic photos in this book were done by Dr. Sarath Vega Gutierrez, using wood from Dr. Robinson's private collection.

Ray Van Court, M.S., helped with the research on wood toxicity for the appendix on food safe woods—a tedious, thankless task that we all deeply appreciate!

Finally, hundreds of members of the Wood Education and Safety Facebook group contributed photos and stories that helped make the book rich and vibrant.

To everyone who helped make *Living with Wood* come alive, thank you!

*For Scout,*
*who introduced me to the world of wooden toys*

# Contents

**INTRODUCTION** . . . . . . . . . . 10

**CHAPTER 1:** Fundamental Wood Anatomy . . . . . . . . . . 12

**CHAPTER 2:** Wood Buzzwords . . . . . . . . . . 20

**CHAPTER 3:** Wood in the Kitchen . . . . . . . . . . 23

**CHAPTER 4:** Wooden Toys . . . . . . . . . . 42

**CHAPTER 5:** Cleaning Wood in the Home . . . . . . . . . . 54

**CHAPTER 6:** Decks, Outdoor Play Structures, and Patio Furniture . . . . . . . . . . 62

**CHAPTER 7:** Quick and Dirty Wood Repair . . . . . . . . . . 66

**CHAPTER 8:** Finishing Wood . . . . . . . . . . 81

**CHAPTER 9:** Coloring Wood . . . . . . . . . . 84

**APPENDIX I:** "Unsafe" Woods for Teething . . . . . . . . . . 92
**APPENDIX II:** List of Sellers of Wooden Toys . . . . . . . . . . 96
**APPENDIX III:** Quick and Dirty Wood ID . . . . . . . . . . 97
**APPENDIX IV:** Next Steps in Reading . . . . . . . . . . 124
**APPENDIX V:** Basic Woodworking Machines, Tools, and Their Applications . . . . . . . . . . 126
**APPENDIX VI:** Wood Science Education Programs in the United States . . . . . . . . . . 137

**GENERAL BIBLIOGRAPHY**

Appendix I References . . . . . . . . . . 140
Index to Wood Species Discussed in This Book . . . . . . . . . . 141
Glossary . . . . . . . . . . 143

# Introduction

## *Why does this book exist?*

Humans have lived and worked with wood for their entire history, whether through tools, fuel, or craft. It is logical then to assume that we would *understand* wood, intrinsically.

But we don't.

Part of the reason may be due to wood's nearly ubiquitous nature across our planet. If a wooden tool broke, we could just go make another. If one tree made you sneeze, you just used a different one. And who cared about wood that was burning? If it made you warm, well, what else was there?

Unfortunately, though wood is most definitely a renewable resource, like any resource, it is under increasingly heavy burdens as the world population increases. Being good stewards of the planet means not only recycling and using *less*, but using what you have in a *smart* way. It means not just using wood, but *understanding* wood, and how it interacts with you, your home, and your family.

The science of wood (very conveniently called "wood science") is not a new area of expertise. Wood scientists (broadly defined) are prevalent in most countries with forests, especially if said countries are still "developing." The field spans every possible area of science, design, and art, from wood chemists (the people who put chemicals in wood to keep it from rotting too fast); wood physicists, architects, and engineers (think buildings, wood composites, etc.); and wood anatomists (how *do* you tell a cedar from an oak?) to stay-at-home parents who make teething toys for their kids, and wood sculptors.

Although this book focuses on common wood uses in the United States, it is worth noting that wood saturates the daily life of almost every human being. The information presented in this book is not meant as a high-end academic reference or textbook. Its only purpose is to offer parents, woodworkers, bored teenagers—whomever—a gateway into the fascinating and surprisingly complicated world of wood. It covers a broad range of topics, from a primer of wood anatomy to detailed instructions for cleaning wood kitchenware. It was written by a scientist, but also a parent, a woodturner, and a person who has been asked one too many questions about wood floor cleaners. It is not meant to supplant a wood science education, but rather to offer a doorway into a discipline most people don't realize exists.

*If you are new to wood, welcome.*

*If you are a woodworker, a toymaker, a boat builder, or an architect, welcome.*

*If you're a parent wanting to move from plastic toys to wood, welcome.*

*If you have wood in your home and just want to take better care of it, double welcome.*

Everyone, this book is for you. It's also for the more than four thousand people who have banded together in the Wood Education and Safety group on Facebook over the past four years, all dedicated to better utilizing this amazing material. These people have built a community of science and understanding that has given information to thousands. It's exciting to finally share our community with the broader world.

*Welcome, friends, to wood.*

# Chapter 1

## FUNDAMENTAL WOOD ANATOMY

### *What is wood?*

At its most basic, wood is a collection of cells, water, sugars, secondary metabolites, and a few other items that form the basis of trees and shrubs. It generally refers to dicots, not monocots (bamboo is a monocot), although an argument could certainly be made either way. The main idea is that it is *fibrous* and *porous*, two features that help explain why wood moves and behaves the way it does.

Generally, wood is separated into two categories: hardwoods (deciduous trees) and softwoods (coniferous trees). The terms "hardwood" and "softwood" have *absolutely nothing to do with the density of the wood*, no matter what the sales reps at big-box home improvement stores tell you. Some hardwoods are very light and easy to dent. Some softwoods are incredibly dense and decay resistant. Trees have a *huge* natural variety. The distinction comes from their anatomy, not their density.

# HARDWOOD

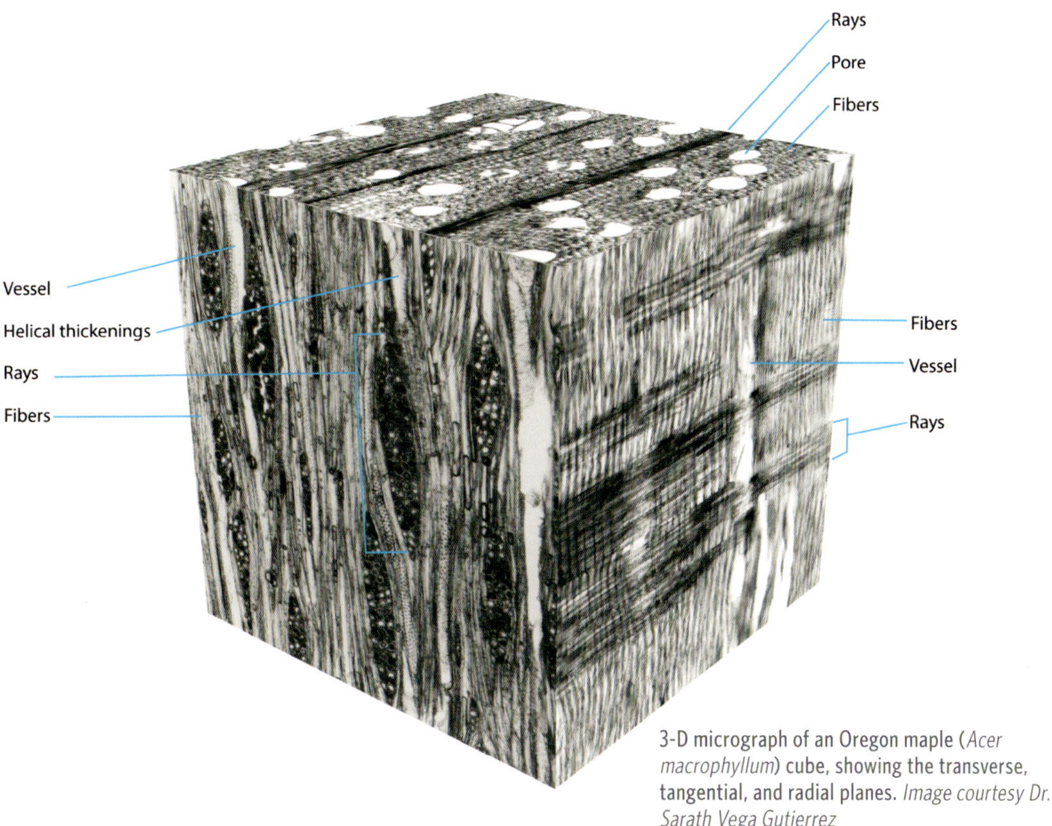


3-D micrograph of an Oregon maple (*Acer macrophyllum*) cube, showing the transverse, tangential, and radial planes. *Image courtesy Dr. Sarath Vega Gutierrez*

Hardwood trees have a bunch of different cell types, as shown on page 97. It is not the purpose of this book to get down into the weeds on anatomical identification, so herein the most "important" cell types have been identified and labeled.

Vessel elements (also called "pores") occur only in hardwoods. If you've ever been told that wood is like a bunch of upright straws, those "straws" are the vessel elements (think smoothie straws, though). They usually run up and down the tree and do a lot of transport.

There are smaller straws in a tree that are called rays, and these run in and out (so at a 90-degree angle to the vessel elements). You can think of these cells like coffee stirrer straws. Rays transport and store things for the tree, depending on the type. The rest of the tree is primarily made up of fibers (sometimes called fiber tracheids, functioning mostly to give mechanical support to the tree), parenchyma (storage cells . . . think of them as the tree's lunch box), and the occasional tyloses, which are actually parenchyma cells that have moved into a vessel element and blocked it. Tyloses don't occur in all species, but you're very familiar with the species that have them. But more on those later.

# SOFTWOOD

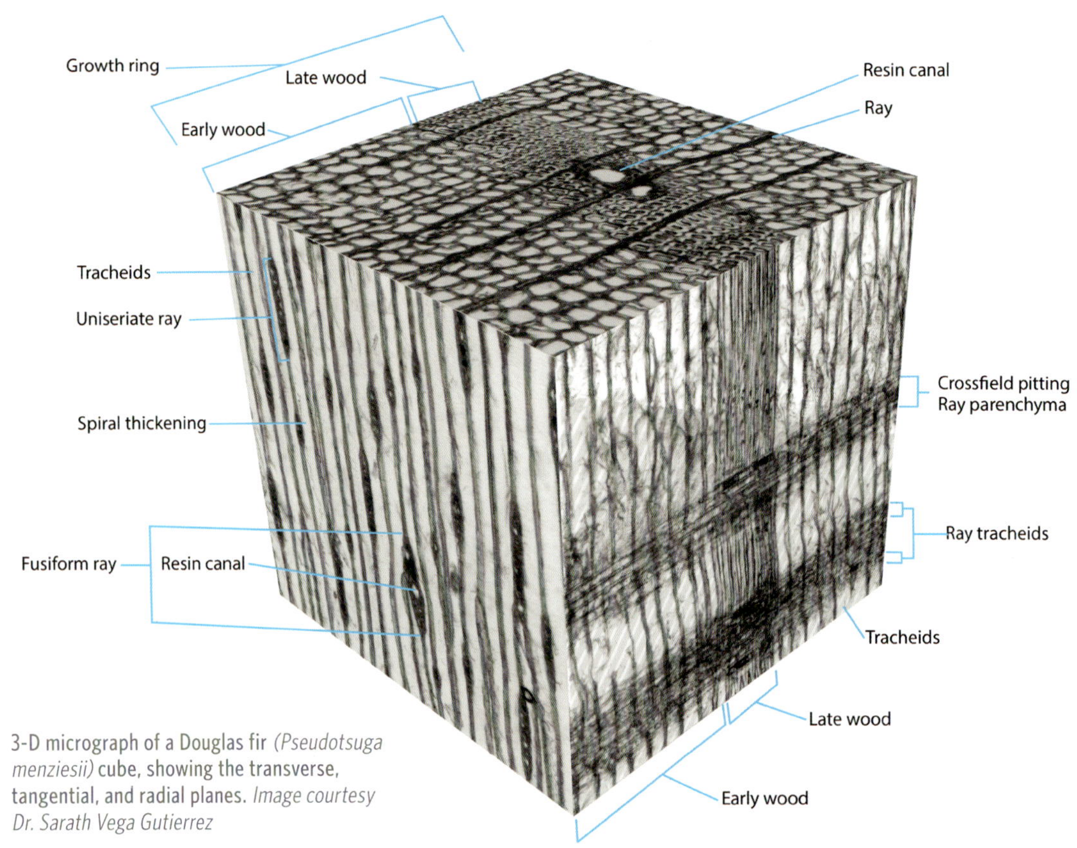


3-D micrograph of a Douglas fir *(Pseudotsuga menziesii)* cube, showing the transverse, tangential, and radial planes. *Image courtesy Dr. Sarath Vega Gutierrez*

There are very few cell types in softwoods. Most softwood trees are almost entirely longitudinal tracheids, which serve as a sort of in-between vessel element and fiber tracheid (meaning they are used both for conduction *and* support). Tracheids are not open at the end like a vessel element, so instead of envisioning them as new straws, think of them as straws that have been chewed up on either end by a bored kid in a restaurant.

**Resin Canals**

*Technically* the resin canal is just a canal, and the cells are the epithelial cells that line the resin canal. But this isn't a technical book, so a little generalization won't kill us.

Softwoods also have rays that function very similarly to hardwood rays and come in several types. (If only one cell wide, they are known as "uniseriate rays"; if they are more than one cell wide, they are multiseriate; and if the ray has a resin canal in it, it is called a "fusiform ray." Wood anatomists like to name things.) There are also longitudinal parenchyma (also found in hardwoods), and a special type of cell known as a resin canal, which secretes that "pitch" you may be familiar with from some conifers (think of pines in particular).

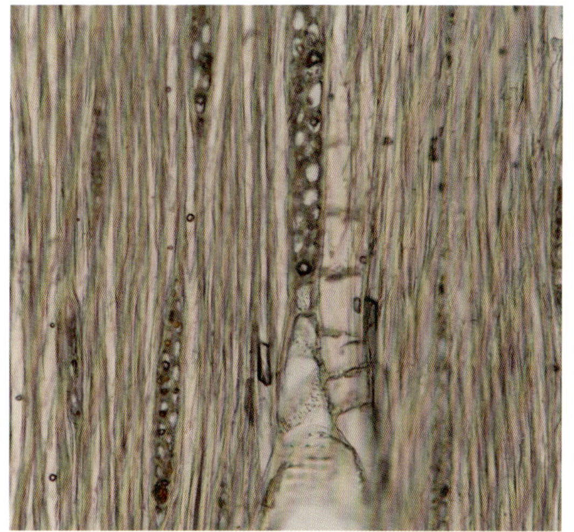

Tangential plane showing empty *parenchyma* (the boxlike cells). The little pea pod shapes are the ends of the rays. *Image courtesy Dr. Sarath Vega Gutierrez*

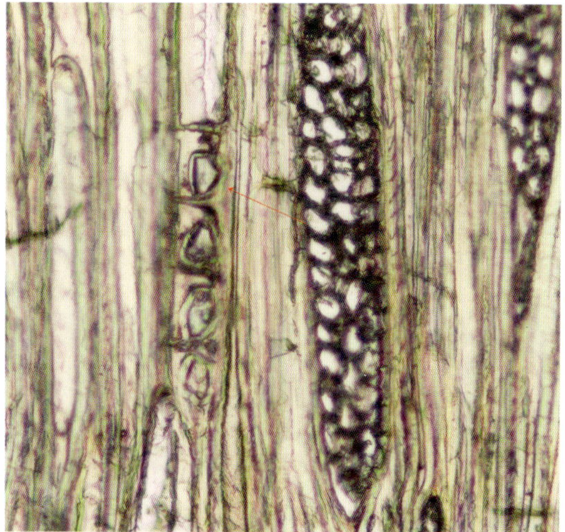

*Parenchyma* cells with "crystals" inside them (box-shaped cells). These types of crystals can be formed through many processes, including by the tree itself, through fungal/tree interactions, etc. The pea pod shapes are ray ends. *Image courtesy Dr. Sarath Vega Gutierrez*

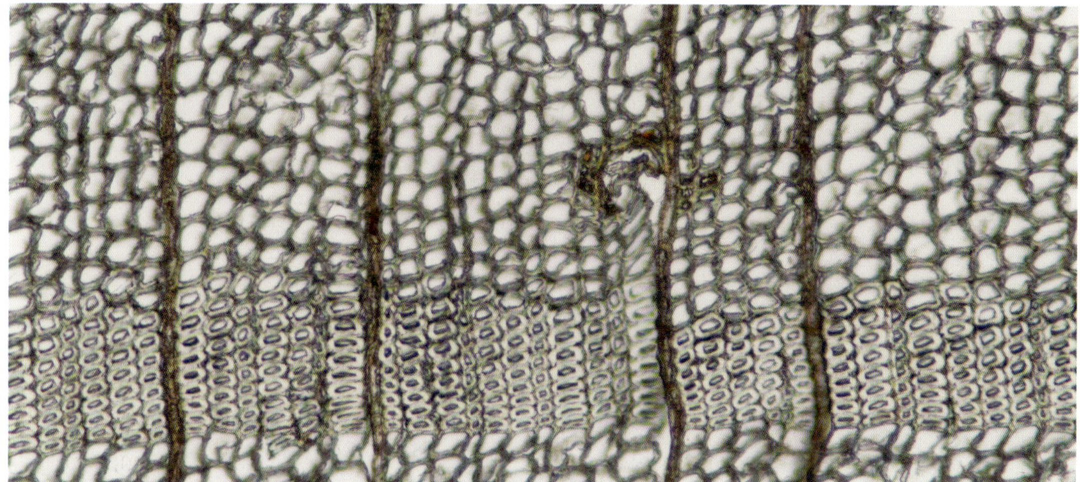

Transverse plane (cross section) of Douglas fir, showing one and a half growth rings. The larger, more open cells are the earlywood, and the smaller, thicker cells are the latewood. One earlywood and one latewood section make one full year of growth, or one "ring." *Image courtesy Dr. Sarath Vega Gutierrez*

Both hardwoods and softwoods, when grown in temperate climates, have *growth rings*, which include one earlywood strip (springwood) and one latewood strip (summerwood) to make one complete ring. When you look at the cross section of wood, the dark lines you normally count as rings to determine the tree's age are just the latewood. It's not technically the whole ring, just the back half of it.

One could argue for the inclusion of a third category, specifically for tropical woods. Many trees that grow in tropical climates such as, say, the Peruvian Amazon rainforest have unusual anatomy that may not follow the "rules" of temperate species. Some anatomical features, such as parenchyma or gum ducts, appear with much more frequency, and the *extractive* content of the trees can be higher. Which leads into a really nice discussion about . . .

## WOOD EXTRACTIVES

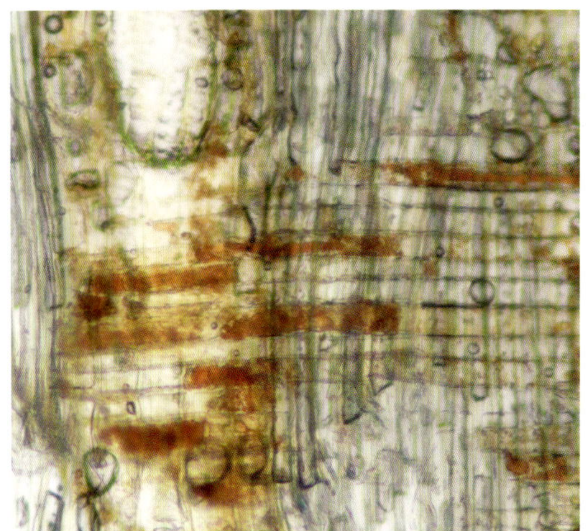

Orange extractives in some procumbent ray cells (radial plane). *Image courtesy Dr. Sarath Vega Gutierrez*

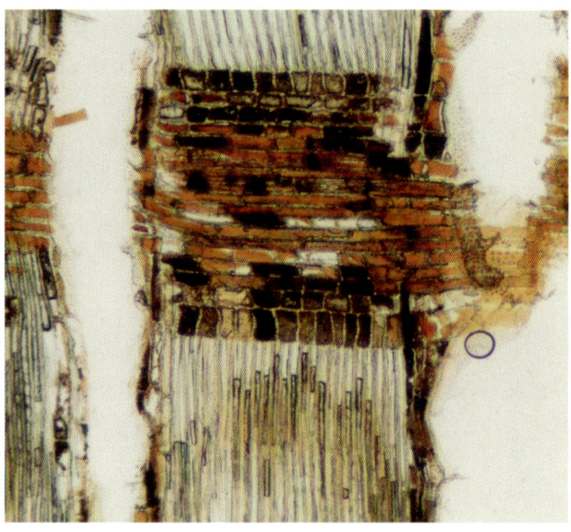

Dark extractives filling almost the entire ray crossing. *Image courtesy Dr. Sarath Vega Gutierrez*

A macro view of gum in vessels (the silver longitudinal lines). *Image courtesy Dr. Sarath Vega Gutierrez*

Corresponding micro view of gum in vessels, in the transverse plane. *Image courtesy Dr. Sarath Vega Gutierrez*

Trees aren't just cells. They're water, too, and sugars, and a bunch of other things, but most importantly for *your* use, they are extractives too. "Extractive" is a nice name for secondary metabolites produced by the tree (secondary because they don't directly relate to growth) that can be "extracted" from the wood through water or other fairly benign solvents (such as alcohols). You can think of extractives as "extra stuff" the tree produces that isn't relevant for its growth but is really useful for such things as defense—defense against other kingdoms, such as bacteria and fungi, but also the animal kingdom. This means you.

Extractives tend to be more numerous in the *heartwood* (the inner part of the tree, often darker) of trees than the *sapwood* (the wood closest to the bark, often lighter). The color difference

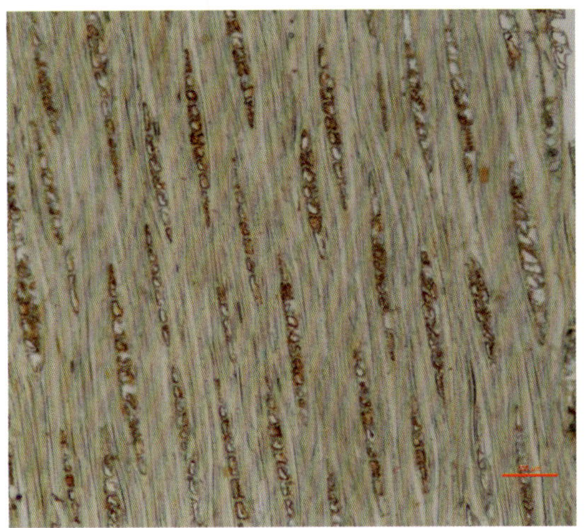

Gum in the rays (orange stuff in the pea pods). Tangential view. *Image courtesy Dr. Sarath Vega Gutierrez*

Raidal viw of gum (the orange stuff in the rectangles) in a series of vessel elements. *Image courtesy Dr. Sarath Vega Gutierrez*

between heartwood and sapwood is often due to extractives. Extractives can make the wood more durable, which benefits the tree in the heartwood, since the heartwood provides a lot of the tree's mechanical support. To be durable, extractives need to be toxic to a wide range of organisms—including insects—in the animal kingdom. And while insects are of course much smaller than humans, that doesn't mean that extractives can't also affect us.

You've almost certainly encountered extractives in your daily life, likely without knowing it. That cedar smell? That's an aromatic extractive. The beautiful dark-brown color of walnut heartwood? Extractives. The flavor and smell of (old-time) root beer came from sassafras trees (mostly the roots) from . . . you guessed it—extractives. Extractives are the active ingredients in many early (and some current) medicines and make up a sizable portion of the essential-oil industry. They've been studied as cancer cures and broad-spectrum toxins, food flavorings, and decorations. They're one of the smallest components of a tree, and yet they have a significant impact on our lives. They're natural, sure, but that doesn't make them safe across the board. For example:

*Sassafras* is one of the compounds used to make traditional root beer. The extractive of interest, safrole, was found to be carcinogenic in rats, and the FDA flat out banned it back in 1960.[1] It was also found to cause liver damage. Commercial root beer from companies such as A&W and Barq's uses a synthetic variety, and the "natural" extracts available on the market generally have safrole removed. Home brews, though, often don't, nor do many of the "natural" and "organic" teas made with sassafras or sarsaparilla. And sassafras wood? Well, it was used as a naturally decay-resistant wood in boats for decades for a reason.

Cedars—whether true, false, or under a common name such as juniper—tend to share a group of aromatic (smelly) extractives that are widespread toxins and irritants to the animal kingdom. Small-animal societies have long warned about the

Wood extractives are also a concern for animals. Chew toys should be made from wood that isn't toxic, but which extractives are problematic will vary by species. In this photo, a lionhead rabbit enjoys a willow wood chew toy, unfinished.

dangers of cedar woodchip bedding,[2] and an unfinished cedar sculpture is credited for making a sizable portion of the FBI sick back in 2016 (although the actual cause of the sickness is still hotly debated).[3] Although a host of extractives are at work across the cedars, cedrol is perhaps the most well studied. While as a *flavor* it is approved by the FDA and does not show any dermal reactions, the smell has a known sedative effect on adult humans[4] and has been found in some studies to actually induce sleep.[5] And while a natural sleep aid may sound amazing at two in the morning, it is important to remember that the size of an adult human—who may feel only drowsy—differs wildly from a dog or small child, within whom results could be much more severe.

Another example where the size of the person makes all the difference is with the extractive camphor, which comes from the camphor laurel tree. The USDA restricts the amount of camphor that a product can have to less than 11 percent because of the horrific effects it can have on children (there's a reason a lot of those chest rubs for cold treatments can't be used on very young children). It is known to cause seizures in children[6] and, in some cases, death, if they are exposed to too much.[7] It can also have adverse effects (though not to the same degree) in adults.

As a final example, the dark color of walnut heartwood comes from an extractive known as juglone. Walnut wood shavings have long been contraindicated for horse bedding due to laminitis. In one study, internal application of juglone caused pulmonary edema, and the toxic effects throughout appeared to build through continued exposure (a very common effect of wood extractives, as any longtime woodworker can tell you).[8] In humans, juglone is under investigation as an anticancer compound due to its well-known and fairly broad-spectrum toxicity.

There are thousands, if not millions, of wood extractives on the planet, many of which have never been studied for human interactions. What can be said about them is that they *generally* play a role in tree defense, which means they *generally* will have some type of effect on the animal kingdom. What that effect might be, or how much of any given compound is needed to affect an adult or a toddler, will vary on the basis of the amount extracted.

And sure, what's a little juglone to a 200-pound man? A breadboard made of walnut may seem like a lovely holiday gift, but a teething ring made of the same wood, for an infant, raises serious questions. Teak is a very durable wood for indoor or outdoor use due to its very high extractive content, but such a high load of extractives inhibits the natural antimicrobial features of, say, a cutting board.

Wood is an amazing, durable, renewable material, but to be used to its fullest potential, it is critical to apply an understanding of wood anatomy and wood chemistry to its use. The following chapters detail some of the most common domestic uses for wood and offer, on the basis of what we understand currently of *wood science*, a guide to best practices for this versatile material.

---

1. A. B. Segelman, F. P. Segelman, J. Karliner,and D. Sofia, "Sassafras and Herb Tea: Potential Health Hazards," *Journal of the American Medical Association* 236, no. 5 (1976): 477.

2. George Flentke, "The Dangers of Softwood Shavings," https://rabbit.org/care/shavings.html (last accessed December 29, 2018).

3. Matt Dixon, "$750K Sculpture Sickened FBI Workers in Miami," *Politico*, December 2, 2016, www.politico.com/states/florida/story/2016/12/the-750k-sculpture-that-hospitalized-fbi-miami-workers-107768 (last accessed December 29, 2018).

4. S. Dayawansa, K. Umeno, H. Takakura, E. Hori, E. Tabuchi, Y. Nagashima, H. Oosu, et al., "Autonomic Responses during Inhalation of Natural Fragrances of 'Cedrol' in Humans," *Autonomic Neuroscience* 108, nos. 1-2 (2003): 79-86.

5. A. Sano, H. Sei, H. Seno, Y. Morita, and H. Moritoki, "Influence of Cedar Essence on Spontaneous Activity and Sleep of Rats and Human Daytime Nap," *Psychiatry and Clinical Neurosciences* 52, no. 2 (1998): 133-35.

6. H. Khine, D. Weiss, N. Graber, R. S. Hoffman, N. Esteban-Cruciani, and J. R. Avner, "A Cluster of Children with Seizures Caused by Camphor Poisoning," *Pediatrics* 123, no. 5 (2009), http://pediatrics.aappublications.org/content/123/5/1269 (last accessed December 29, 2018).

7. J. N. Love, M. Sammon, and J. Smereck, "Are One or Two Dangerous? Camphor Exposure in Toddlers," *Journal of Emergency Medicine* 27, no. 1 (2004): 49-54.

8. R. G. True and J. E. Lowe, "Induced Juglone Toxicosis in Ponies and Horses," *American Journal of Veterinary Research* 41, no. 6 (1980): 944-45.

# Chapter 2

## WOOD BUZZWORDS

### *Can wood be organic? Can it be "free range"?*

Sure, wood can be organic. It can be grown in a forest without pesticides or human intervention. It can be "free range" too, grown in the wild and not on plantations or under intensive management.

But do these words really mean the same thing when applied to trees? Let's take a minute to break down the three most popular wood buzzwords and understand what they're *really* telling us.

## SUSTAINABLE

If you were going to cling to a buzzword surrounding wood and trees, sustainable should be that word. A great deal of tropical wood imported to the US is not legally logged, causing massive deforestation issues in other countries. And while the *word* "sustainable" means little within itself, there are groups such as the FSC (Forest Stewardship Council) that work under a directive of sustainability.

The little tree icon with the FSC Certified stamp is something to look for on wood brought in from outside the US. Such a stamp indicates that the wood was taken from second- or third-growth forests and then replanted, or within other allowed norms of the organization. With FSC certification, there is a known chain of custody for *each tree*, meaning you could track any FSC tree back to its original plot of land, thereby ensuring that it was legally cut. The FSC uses certified forest management plans and makes attempts to regulate controlled woods.

**Wood and wood products play critical roles in our lives, from the obvious to the not so obvious.** *Image courtesy Candace Hermann: https://stayingincourage.com*

It's important to note that not all groups that call themselves sustainable are. FSC is generally regarded as the most reliable and sustainable of the bunch.

## ORGANIC

Technically, all wood is organic. Grown on a plantation, in a lab, or in a rainforest, a tree is still wood, no matter what. There's some concern about plantation trees being sprayed with pesticides, but this is a misunderstanding. In the US, federal lands are not sprayed for reforestation. They are sprayed only for invasive species, meaning the *trees* aren't sprayed, just the stuff that's trying to kill them.[*] Even in intensively managed forest lands, the spraying is generally done three years out within a forty-year rotation, and, again, the *understory* is sprayed, not the trees.

If you want to be double, triply sure that pesticides have not come anywhere close to your wood, the FSC label, again, is what you want, since FSC-certified wood can't have had any pesticide put on it, and also has restrictions on plantation establishment.

It's important to remember that while concepts of free range and organic may seem to go hand in hand in agriculture, they're opposing in the forestry industry. You can't have it both ways. Plantations are intensively managed, generally, and the understory may be sprayed. But the alternative is cutting down old-growth forests and destroying critical wildlife habitat, as well as the homes of the few remaining indigenous tribes. Trees from old-growth forests also tend to have much-higher extractive contents, and those chemicals don't have the same type of testing that many of the synthetics in pesticides do.

It's not about choosing right or wrong here. It's just about making an educated decision based on the information available.

**Concepts of "free range"—works great for bunnies, not so much for wood.** *Image courtesy Kristen Pickens of Two Raccoon Hollow, using cherry, maple, walnut, and poplar wood*

## FREE RANGE

I assume this means "not a plantation tree" when referring to wood. And sure, it'd be nice if trees didn't have to be grown in monocultures just to be cut and turned into utility poles. It's really easy to draw parallels to caged animals on factory farms. The issue is as previously discussed: a "free range" tree would be an old-growth tree, or a tree in a natural regeneration stand. And to let these types of forests grow, to create critical wildlife habitat, and to preserve some of the oldest living organisms on the planet, we *have* to use plantation trees.

You can see trees, forests, and forestry as agriculture (as they do in many countries—Chile in particular), with trees as the crop, or you can look at it all as a wild ecosystem that needs saving, but in reality humans live forever suspended between these two ideas. We *need* wood to live. We have always needed wood to live. But so too do many other organisms.

The trick, really, is not so much black and white, plantation or old growth. It's balance, and understanding, and, well, *wood science*.

* Region 8 sprays a bit, but minimally.

# Chapter 3
## WOOD IN THE KITCHEN

Wood is both *hygroscopic* (can absorb water from the air) and *hydroscopic* (can absorb water generally from, say, being washed). The water in wood tends to move over time, so that the wood is uniformly wet, instead of just wet on the surface after you wipe it down.

These are very important concepts because they relate to how wood functions and how you use it in your kitchen. Most kitchens, whether wood friendly or not, tend to have at least one wooden cooking spoon and cutting board lying around. Others have full-on wood countertops, wood mixing bowls, and wood serving dishes. Humans have used wood in their kitchens for as long as kitchens have existed, but with so many alternative products on the market, is there really a benefit anymore to wood?

And, you know, what about all that cracking? What about *mold*?

Let's discuss.

Glue-up rubberwood cutting board that was run through a dishwasher. The heat and water swelled the wood and popped the glue seams. The wood then warped from the differential pressures during drying. *Image courtesy Carol Johnson*

## WOOD AND MICROORGANISMS

Water exists in two states in wood: *bound* and *free*. Bound water is water bound to the cells. Think of it more as being tied up and therefore hard to remove. Free water is liquid water and exists only once all the sites that *could* hold bound water are full. Free water tends to exist mostly in cell *lumens*, which are the empty spaces in the middle of cells (the hollow of your straw).

The *fiber saturation point* is sort of a magical place in wood where all the binding sites for water are full, but there's not yet any liquid water. It's a good phrase to know, because scientists often refer to the fiber saturation point in reference to wood drying and wood decay, because it's all linked together. The fiber saturation point for wood, generally, is around 30 percent. It is much *lower* for high-extractive-content woods, such as ipe, because their binding sites are already filled up with other things. It can also be higher for some woods that have more binding sites available.

Many (but not all) microorganisms prefer wood to be *above* the fiber saturation point, or somewhere close. Most dry wood in the home hangs out around 12 percent moisture content (MC), depending on what sort of heating / air conditioning and exterior climate exist. Interior wood in Arizona, for instance, where the *relative humidity* is generally lower, will have a lower MC than wood in Hawaii.

We dry wood and move it significantly below the fiber saturation point for a number of reasons, but a big one is to prevent decay. "Dry" wood (for this book we will define that place as below about

You don't need water and heat to damage wood. Heat alone works just fine. In this image, a hot frying pan is permanently stuck to a wooden cutting board that was used as a trivet. Excessive heat can actually melt certain parts of wood, forming a natural and very strong glue. *Image courtesy Valerie Carroll*

Color left behind by mold fungi. This cutting board was dried propped, but with the end on a mat that did not readily evaporate water. Due to this, the wood stayed in water contact at the bottom, and fungi began to colonize. A wash removed the fungi, but the stain remains. *Image courtesy Letty Camire*

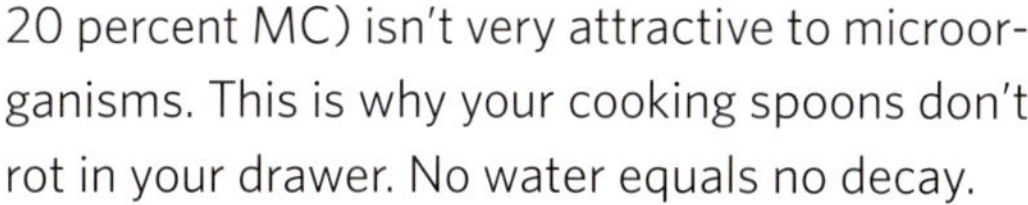

20 percent MC) isn't very attractive to microorganisms. This is why your cooking spoons don't rot in your drawer. No water equals no decay.

So when you *cook* with your cooking spoon, what happens? You introduce liquid water to the system. Your *sorption sites* fill up with water, and then free water begins to fill the vessels. You are also introducing heat into the wood—somewhat similar to a steam-bending process—which weakens the hydrogen bonds and makes your wood a bit flexible. Water moves. Water binds. Water absorbs.

Your wood is now a target.

The great thing is, microorganisms take some time to really establish. They're in the air all the time, of course, and there are plenty on your cooking spoon well before you use it. But they aren't reproducing en masse because of a lack of water. Give them water and *BOOM*! They're off! But not overnight. This is why it's so critical to dry your wood immediately after use, and to do so propped, not flat on a surface. Get the wood below that magical MC and the microorganisms will not be able to continue growing.

Of course, it's not *quite* as cut and dried as that. If you lay your cooking spoon flat on the counter to dry, only one side will dry well. Wood likes to equilibrate its water, but it takes some time to do so. The spoon may even feel dry on both sides, but since there has been so much less air flow to the backside, there is still quite a bit of water in the spoon. Many a spoon has been molded due to being put back in a drawer—a nice, dark, warm drawer in a kitchen—without having been properly dried. Drawers don't get good airflow. Things don't dry in drawers. Wet wood needs to be thoroughly dried before being put away, and that includes giving the wood time to lose that free water it accumulated in the cooking process. Failure to do so means introducing a lovely resort to your microorganism communities and losing your favorite cooking spoon.

Cooking utensils with normal wear after two years of use. All are beech wood, and all have been in contact with boiling water. Those used in greasy applications, such as bacon cooking, have a darker color due to the absorbed oil. All these utensils are in fine shape and have at least a decade of good use left in them, if not more. *Image courtesy Letty Camire*

## WOOD AND ITS ANTIMICROBIAL PROPERTIES

*(also known as OMG, WOOD IS POROUS—DON'T USE IT TO CUT MEAT! and other falsehoods)*

So if wood is so squirrelly with water, why use it? It shrinks, it swells, it cracks, it molds—what a pain! But it's those exact properties that make wood so ideal for kitchen and toy use. Because wood evens the water out when it gets wet, that often means that dry wood that is used to cook with ends up wet on the outside but still has a dry core. To even out, water evaporates from the surface but is also pulled into the wood, bringing the surface microbes with it. *Boom*—clean wood! And all you had to do was be patient.* And yes, the microbes stay inside the wood and don't migrate back out. They eventually die in their little wood cell tombs.

What is less clear, unfortunately, is the effect of wood finishes on this antimicrobial action. Logic would dictate that if you put an oil finish on your wood (like many places suggest, as a conditioning agent to keep your wood looking "shiny" and to prevent cracking from the constant swelling-and-shrinking process), it would eventually block the cells up and the wood would no longer work properly.

Ak et al. found that mineral oil did not significantly affect the functionality of a wooden cutting board, but mineral oil is a nonhardening oil (meaning in theory you could keep moving it in and out of wood with soapy water washes) and no longer the most commonly used cutting-board conditioner (with many consumers now using oil-wax emulsions).

So what effect *does* a finish or conditioner have on wood? First, let's take a closer look at cutting boards.

* This effect has been studied ad nauseam. It takes about twelve hours, more or less, for wood to finish its self-imposed cleaning cycle, but after that the bacteria load is reduced by about 98 percent.[1] Cliver[2] put together a decent review paper on the subject if you want to get into the weeds.

## CUTTING BOARDS

Cutting boards, for the purposes of this book, are defined as boards made from wood (not monocots such as bamboo) that are used for the cutting of human food. They can be a single slab of wood, an end-grain glue-up, or strips of several different woods glued together. They may be finished or unfinished, may have been passed down through generations or installed in a kitchen, or may be an off cut of a board picked up at a box store.

Some people use their boards as cheese boards, breadboards, or serving trays. Some use them only to cut vegetables. Some use them for meat. Some people wash them in the dishwasher; some wash them with just a wet rag. Some oil them after each use; some never oil. There are hundreds if not thousands of ways to interact with a wooden cutting board. None are *bad* ways, but some are more useful and take better advantage of the properties of wood than others. Let's take a look at a few of the most common.

Wood end-grain glue-up cutting board (larch) that takes advantage of the primary cell direction in wood to move microorganisms. *Image courtesy Don McClure*

Solid-slab maple cutting board in service for four years, showing normal wear. Solid-slab wooden cutting boards are used to minimize glue, which can block void space in wood and inhibit antimicrobial function. Solid-slab boards can be more prone to minor cracking and warping if cared for improperly. *Image courtesy Letty Camire*

Solid-slab cutting board containing heartwood and spalted sapwood. Cutting boards such as this are nice decorative pieces but may not function as well, since the density differences between the heartwood (dense) and the sapwood (less dense) may affect cutting. Heartwood also has a higher extractive content than sapwood, which means it absorbs less water, and therefore also fewer bacteria from the surface. The spalting, though caused by fungi, does not affect the usefulness of the board. *Image courtesy Jim Robin, by Judith Jayne Tanzman Photography*

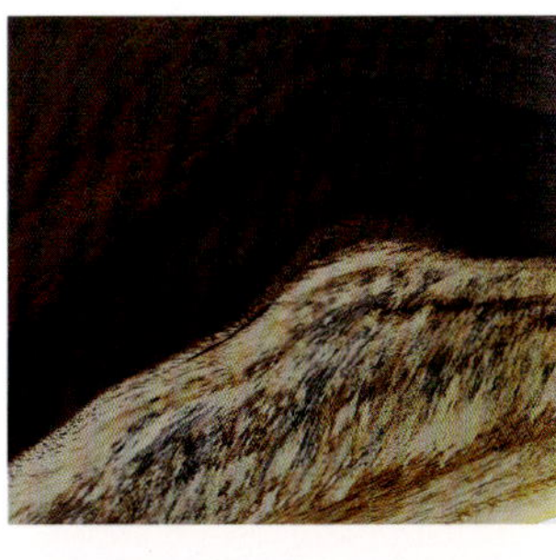

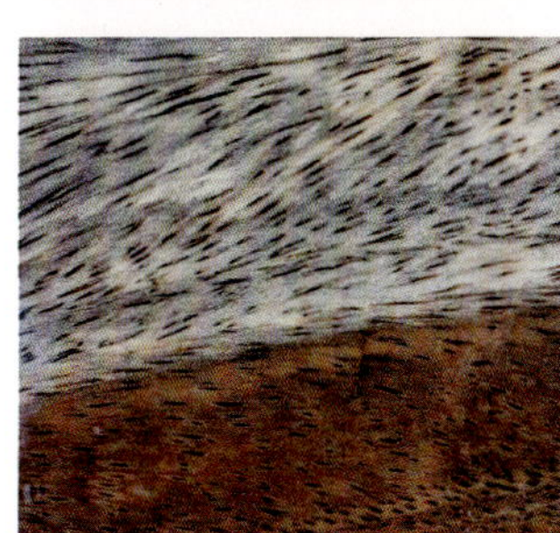

### Orientation and Glue-Ups

In theory, an end-grain cutting board (where all the straws are oriented up and down) would move bacteria more quickly, since there is less *stuff* in the way (in terms of wood cell walls). In practice, end-grain boards tend to be glue-ups made from dozens of end-grain pieces. The glue used to hold the pieces together has also gotten into parts of the vessels, where your water and bacteria want to go. Does the glue cancel out the extra movement from the end grain? I doubt anyone has studied it, but it's something to think about. Academically, the more glue in a board, the less space there is for water to move, so potentially this could affect how fast and how much water (and bacteria) is pulled into the wood. Potentially.

### Wood Species

Peer-reviewed studies have found very little difference between wood species in terms of their ability to move and trap microbes (see citations provided earlier). There are bound to be *some* differences, because higher extractive woods have more stuff in them, and therefore potentially less space for water. But all wood is still porous, and still hygroscopic. Whether or not the wood is food safe is an issue, and consumers should always be aware of the wood species they have purchased, since not all wood is fit to be in contact with food—especially wet food.

### Washing

Studies done by researchers at UC Davis found no significant difference in surface bacteria when boards were washed with soapy water versus just water, which makes sense, since it's the water doing the action, not so much the soap. A number of studies note that putting cutting boards in washing machines is not a good idea, since the pressure from the machine sends the bacteria all over everything else in the machine. And if the wood hasn't had time to move the microbes down, they're still on the surface and get spewed around, same as with plastic.

Really though, putting *any* wood in a washing machine is an awful idea. You don't need to saturate your wood to clean it, and *steam*, or even just hot water, is how you *steam-bend wood*. It brings your MC way, WAY up, loosens those hydrogen bonds, and allows your wood to bend . . . and also to become more brittle. Changing the MC of wood (up to the fiber saturation point) changes its dimensions, which means if you have a glue-up, your board has a high likelihood of breaking its glue seams and falling apart. And if you don't dry the board perfectly, it will warp, potentially leaving it unusable.

Keep wood out of dishwashers. Should wood ever be accidentally put in a dishwasher and it comes out intact, leave it propped to dry for around five days before putting it away, to make sure all the water has moved where it needs to. If a cutting board in particular is placed in the dishwasher and warps upon drying, it can be flattened using clamps. To do this, run the board through another dishwashing cycle. Remove it immediately after the wash—before the dry cycle—and using C-clamps and bar clamps (and flattening boards with some parchment paper in between), clamp the board down. You won't be able to flatten it completely in the first round, but much like stretching a bow, you can tighten the clamps a bit every few hours. If the wood groans too much or you think it might crack, you can always steam the wood to reintroduce enough hot water to ease the process. Remember: wood dries first from the *outside*, causing case hardening of the shell, then pulls water from the *inside*. If you don't want your wood to crack during this process, you need to keep the outside a bit wet to ease the stress.

If there really is a need to sanitize—truly sanitize your board—you can do so in a microwave. Just don't bake it for longer than about fifteen seconds or so at a go. If you flash *all* the water off and keep getting the wood hot, it can catch on fire.

### Drying

Prop to dry. Always. And leave it out, in the air, for at *least* two days. Give the wood time to do its job.

### Conditioning

Many people believe that wooden cutting boards must be conditioned (by which they usually mean a finish placed on them) to function and to keep from cracking.

Wood cracks due to changes in moisture content. Wooden cutting boards generally crack because they have been mishandled—usually due to being left in dishwater and allowed to absorb more water. Yes, oiling your cutting board will help prevent cracking, but at a cost. Any time you add something onto wood, you are limiting its ability to move water, which has a direct effect on wood's amazing antimicrobial nature.

Since there are no readily available tests on the effects of oil on cutting boards, one is presented here. Two nonpathogenic bacterial strains were used: one of salmonella and one of listeria. Four wood species were used: red oak (*Quercus rubra*), white oak (*Quercus garryana*), sugar maple (*Acer saccharum*), and European beech (*Fagus sylvatica*). Two oils were used: mineral oil (a nonhardening oil) and refined linseed oil (a hardening oil). Two different coating amounts were used: one coat and five coats. Everything was done in replicates of four, for statistical rigor.

The tiny cutting boards (just wood squares) were cut to fit our Petri plates, which had a bacterial medium in them. The wood was inoculated with the bacteria, then stamped onto the plate. The plates were allowed to incubate for different times (right after the inoculation solution dried [T0], after one hour [T1], and after twenty-four hours [T24]), then the colonies were counted. No plates were double stamped.

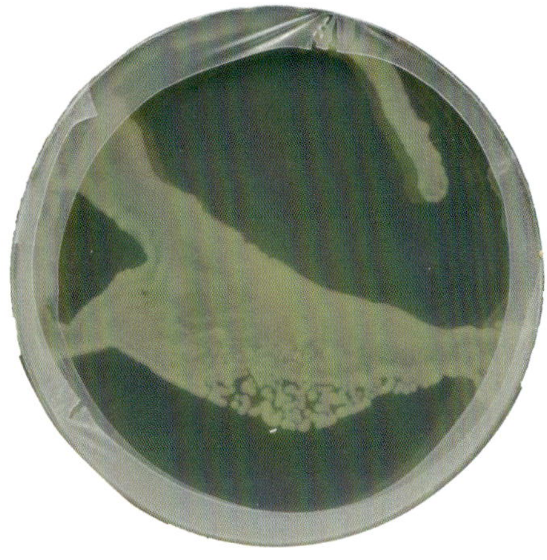

SALMONELLA

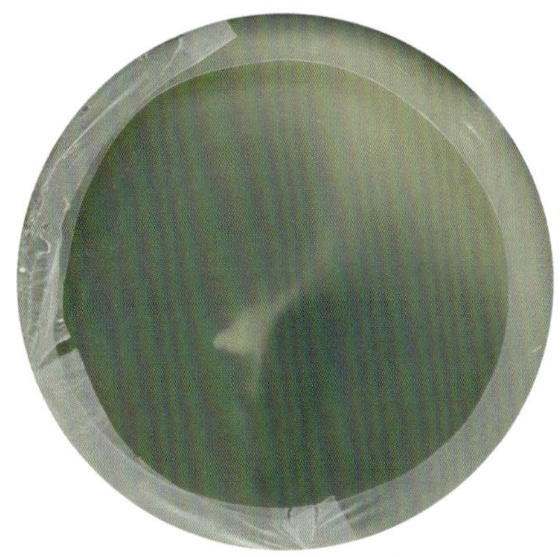

LISTERIA

First, we showed that the bacteria were alive and well.

*All images in this section courtesy Patricia Vega Gutierrez*

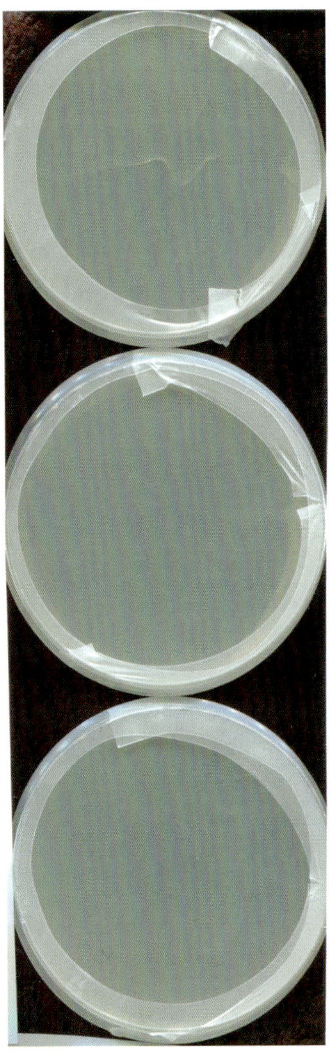

Then we showed that the oils were sterile (and didn't have bacteria growing in them).

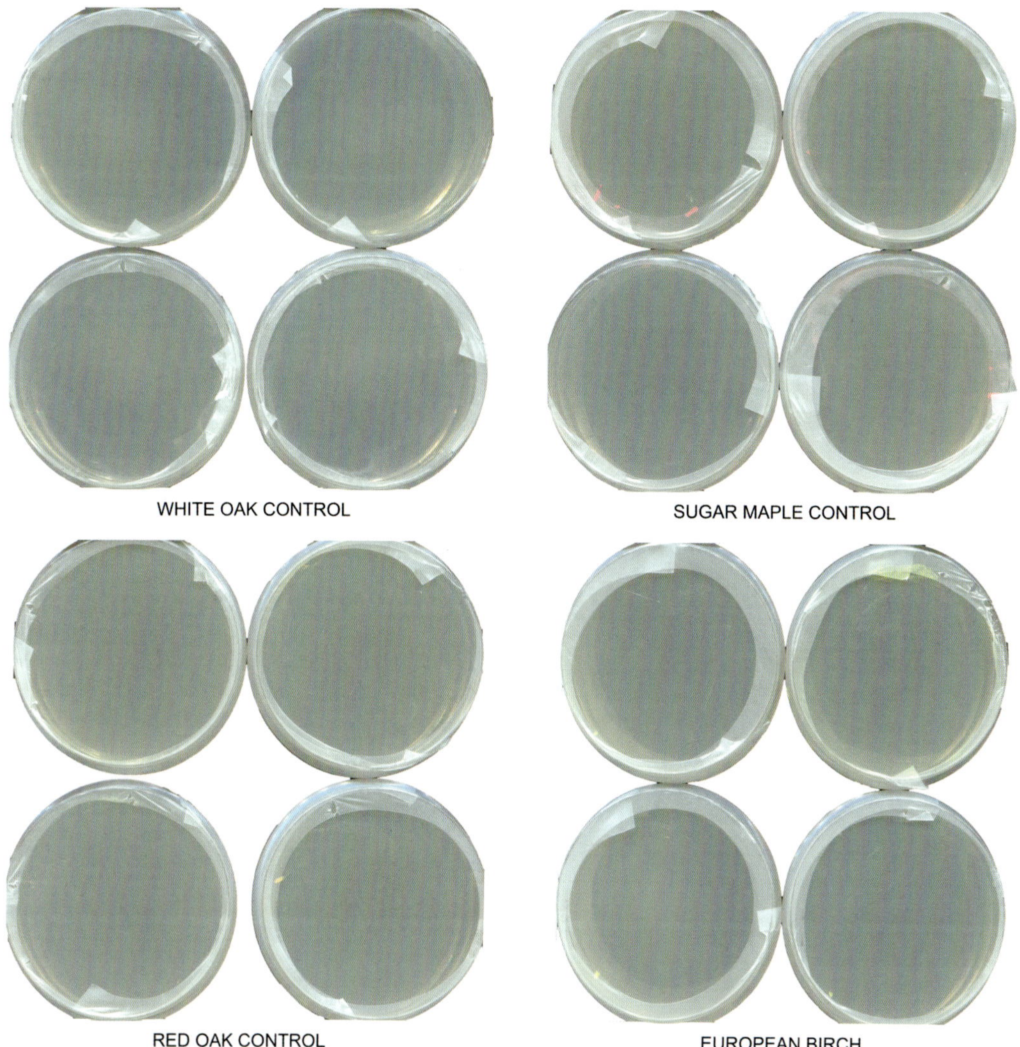


Then we showed that the sterilization process we used on the wood killed *everything*.

The T0 images showed us that bacteria were successfully applied to the cutting boards. Some issues were encountered with the five coats of oil, since the bacteria were applied in a water suspension, and it was hard to keep the water on the wood in the amount required. As such, the five-coat boards had some of their bacteria slip off the side—and so started with fewer bacteria to begin with. As a result, the five coated boards were left off further analysis, since they basically sealed the wood (the first time a knife scored the board, however, that protection would be lost).

Salmonella died within twenty-four hours and thus was gone from all boards, regardless of coating (salmonella is a short-lived bacterium). Listeria was much more aggressive. Robust growth was seen on the control and coated wood.

Even listeria didn't fare well at twenty-four hours and began to die off. Hence, even the coated boards had significantly reduced bacterial amounts at twenty-four hours. While the data from the twenty-four-hour test were included in the statistics, the images aren't shown here (how many Petri plates does anyone want to look at, really?).

A reminder when looking at these images—all are *unwashed* and just meant to show how well different woods with different coatings move bacteria. Washing was not tested.

## European beech

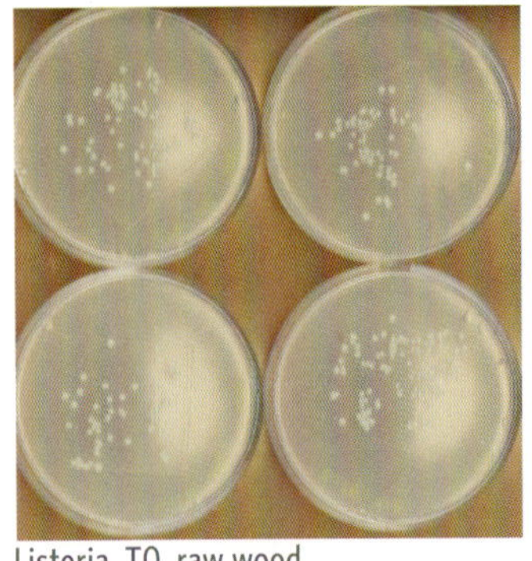
Listeria, T0, raw wood

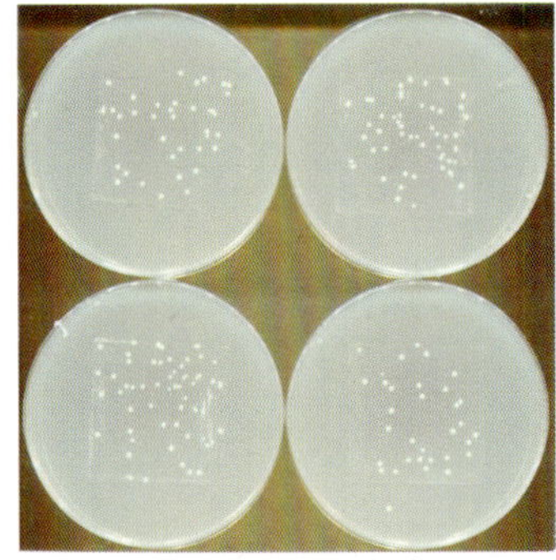
Listeria, T0, 1 coat mineral oil

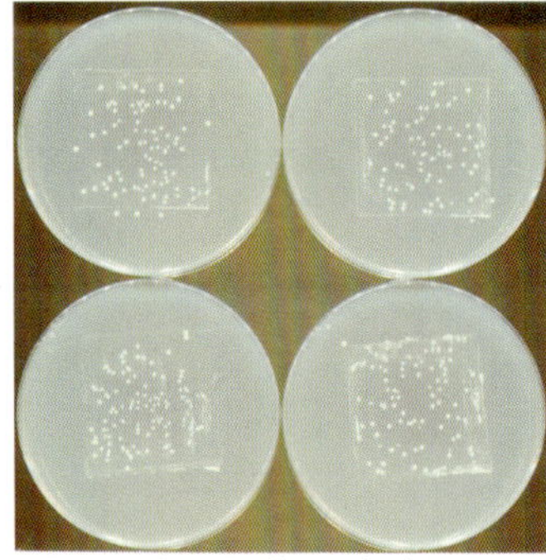
Listeria, T0, 1 coat linseed oil

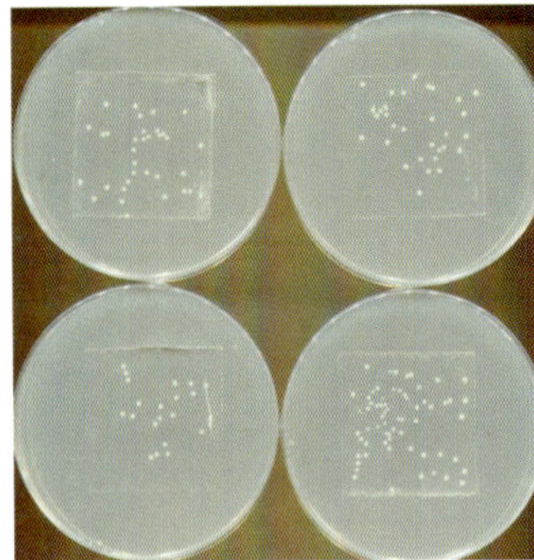
Listeria, T1, raw wood

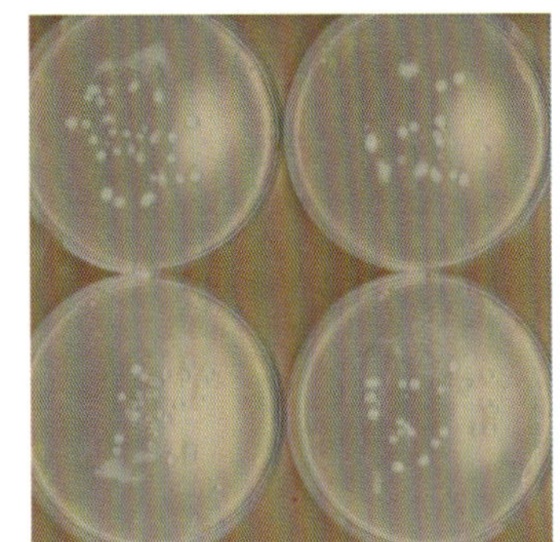
Listeria, T1, 1 coat mineral oil

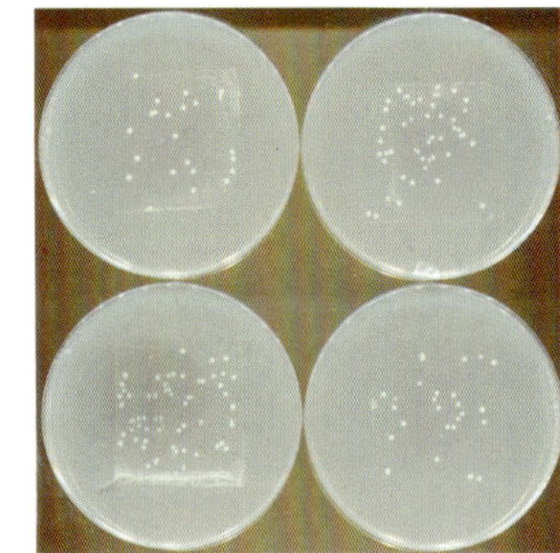
Listeria, T1, 1 coat linseed oil

## red oak

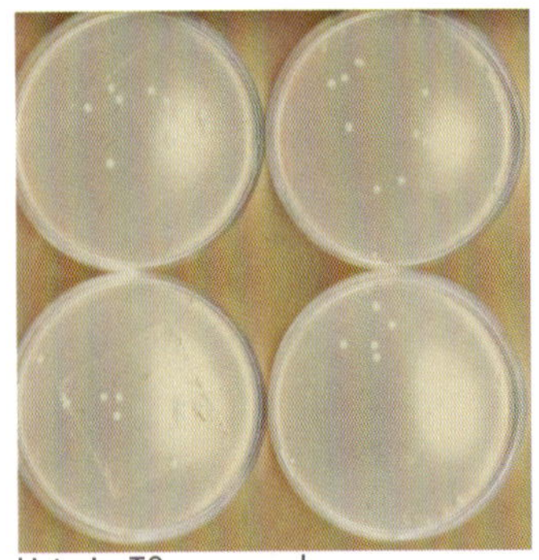
Listeria, T0, raw wood

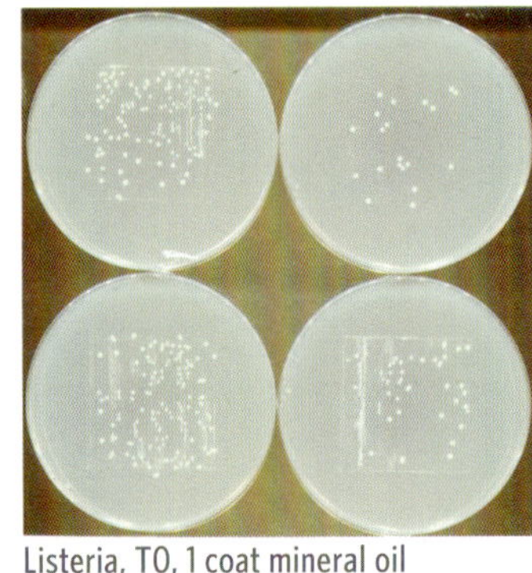
Listeria, T0, 1 coat mineral oil

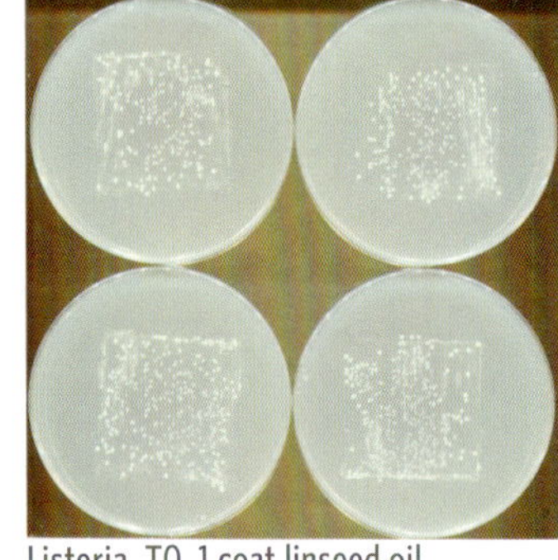
Listeria, T0, 1 coat linseed oil

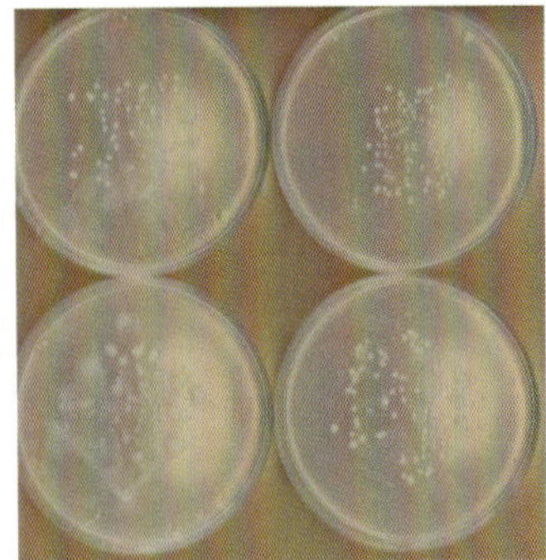
Listeria, T1, raw wood

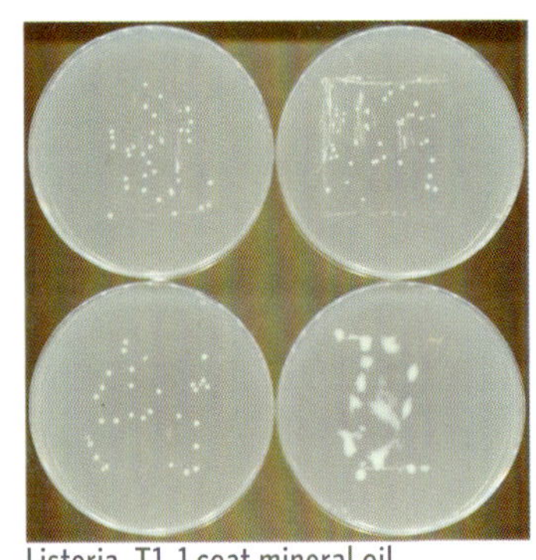
Listeria, T1, 1 coat mineral oil

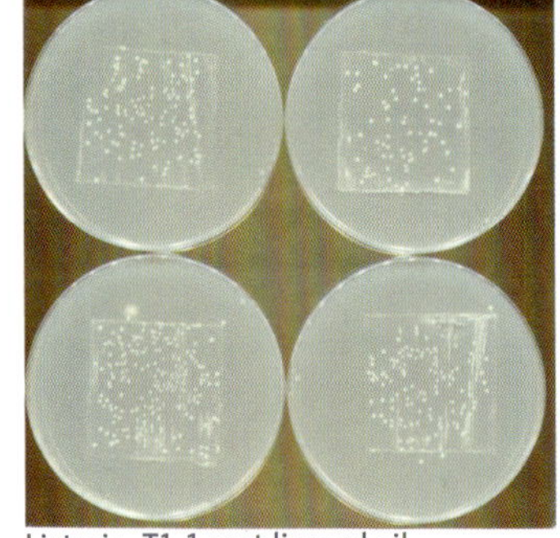
Listeria, T1, 1 coat linseed oil

## white oak

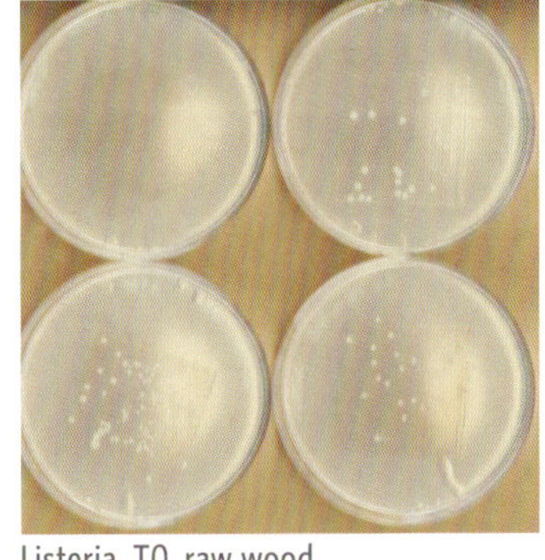
Listeria, T0, raw wood

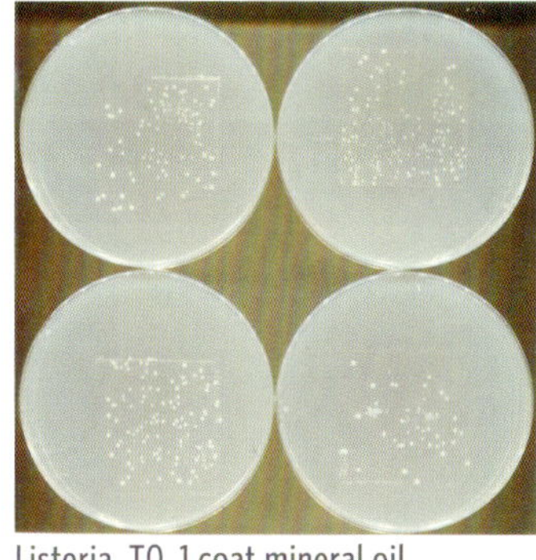
Listeria, T0, 1 coat mineral oil

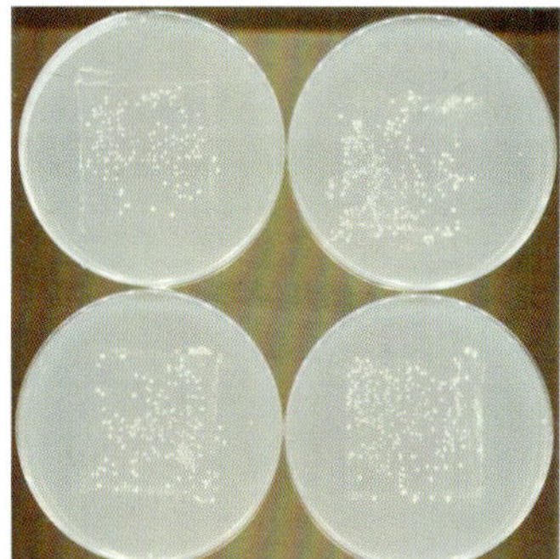
Listeria, T0, 1 coat linseed oil

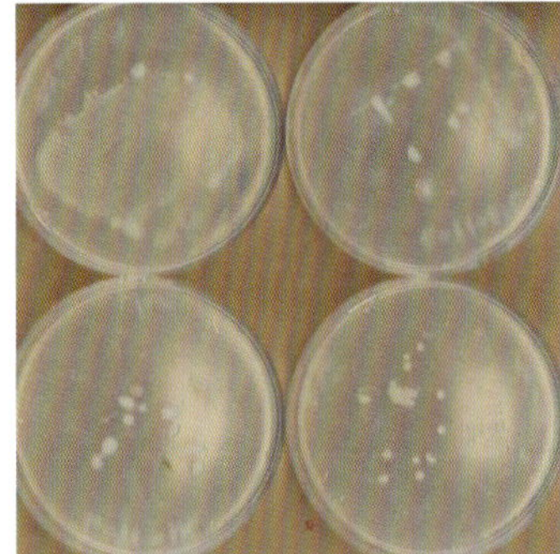
Listeria, T1, raw wood

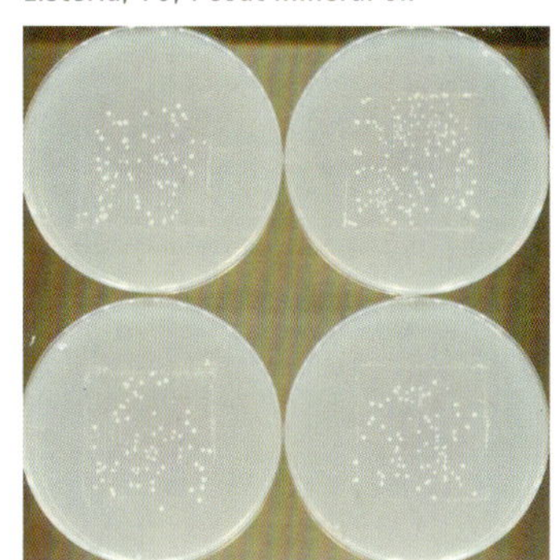
Listeria, T1, 1 coat mineral oil

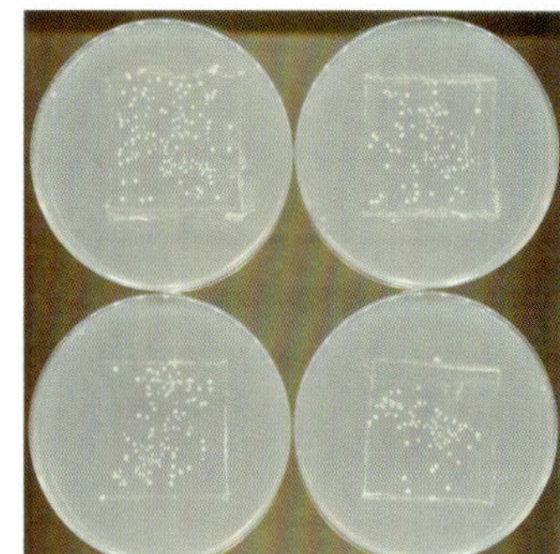
Listeria, T1, 1 coat linseed oil

## sugar maple

Listeria, T0, raw wood

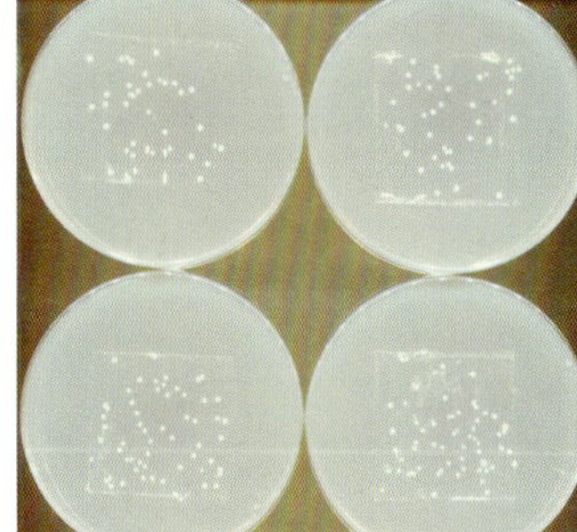
Listeria, T0, 1 coat mineral oil

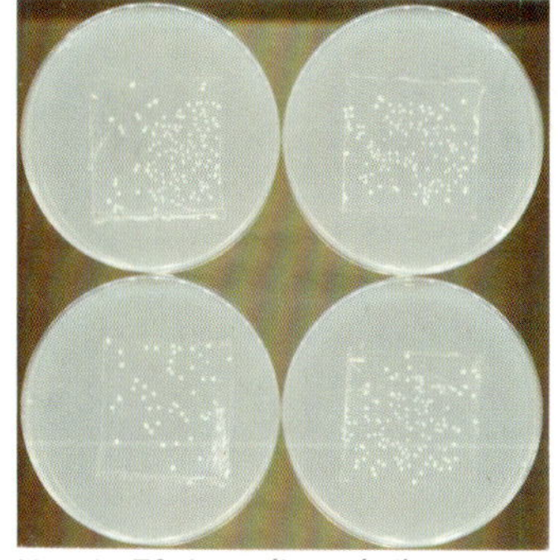
Listeria, T0, 1 coat linseed oil

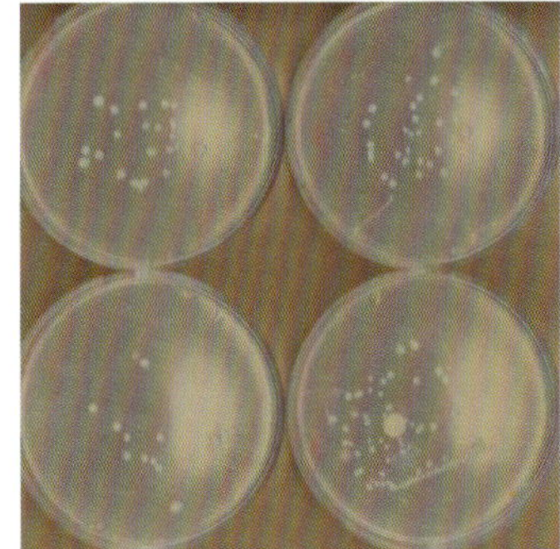
Listeria, T1, raw wood

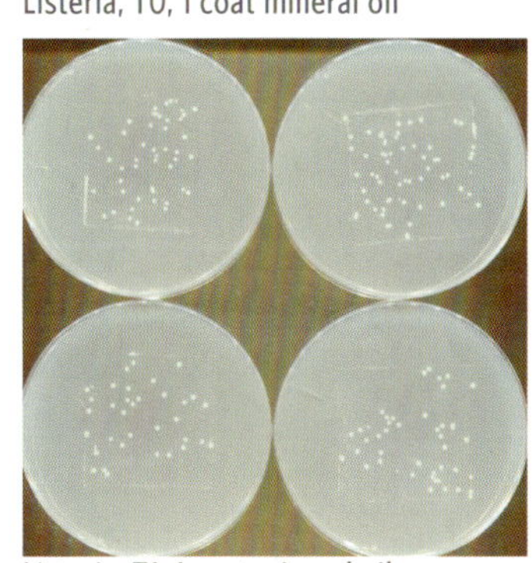
Listeria, T1, 1 coat mineral oil

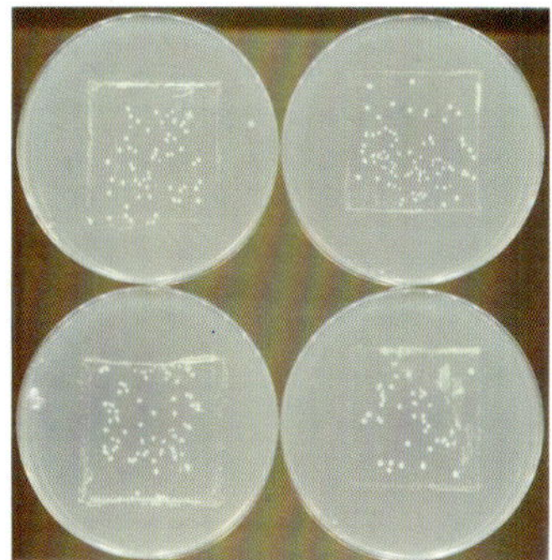
Listeria, T1, 1 coat linseed oil

### The Statistics

Interpretation is fine, but statistics are what matter. The data herein were run on a mixed model to reduce variation and were interpreted with a Tukey HSD (yes, we have a scientific publication forthcoming). Here are the statistical takeaways:

- Coated samples at time 0 (listeria) had the highest amount of colonies (coating blocked the absorption of bacteria).
- Noncoated samples at time 0 (listeria) had similar numbers of colonies as coated samples at one hour (it takes one hour for bacterial colonies to reduce on a coated sample to the amount that occur on a noncoated sample right after use—you could look at this sort of like uncoated boards work twice as fast as coated boards).
- White oak at twenty-four hours incubation with no coating had the least amount of bacteria (not statistically significant, but interesting!).
- Red oak and white oak with linseed oil at time 0 had the most bacteria.
- Linseed oil (generally, and varied by wood species and time) promoted more bacterial growth than mineral oil (likely because it is a hardening oil and limited antimicrobial action).
- At twenty-four hours, everything started evening out as the listeria started to die.

### The Takeaways

Don't oil your wooden cutting boards. If you must oil, use the board *once* and then wash and let it sit for at least twenty-four hours before using again. All oils inhibit antimicrobial action. Wood species (ring porous versus diffuse porous, or tyloses versus none) has very little if any effect.

As an additional note, some people have difficulties sourcing unfinished cutting boards or want to transition their conditioned board to an unconditioned one. If the board was conditioned with mineral oil, this is a fairly simple process and involves just washing the board with soap and water for several months as the oil works its way out.

For boards conditioned with waxes or hardening oils, the finish can rarely be undone. If the board is thick, sanding about half an inch off either side would take care of most of the oil. Most commercial cutting boards do not have this much material to play with. If a board has a history of conditioning and there is a desire not to continue the process, it is possible to simply stop oiling the board. While this will not make the already-applied oil disappear, it will stop further buildup.

Walnut serving spoon not intended for wet contact. *Image courtesy Kristen Pickens*

## COOKING SPOONS

Generally, care for cooking spoons is the same as cutting boards: no dishwasher, and prop to dry for well past when it feels dry to you. Of all the items in the kitchen, it is perhaps most important to consider wood species in your cooking spoons. Since extractives can move readily from some wood species in water—especially *hot* water—using a wood species with dangerous water-soluble extractives as a cooking spoon is concerning. Like making stock from bones, hot water pulls extractives from your wood each time the wood is submerged. And the last thing you want in your pasta water is something that could hurt your family. For a list of woods not to use in food or mouth contact, see appendix I. For a quick and dirty wood identification, see appendix III.

Beech cooking spoon, in service for two years, where the handle has been used to stir dye for clothes. *Image courtesy Letty Camire*

Poplar finished with walnut oil. *Image courtesy Kristen Pickens*

Cherry spatula finished with walnut oil. In use for two years. Wood with oil finishes will show less damage over time than unfinished wood, but the finish can affect the antimicrobial properties. *Image courtesy Kristen Pickens*

Beech cooking spoon, in service for two years, showing normal wear and some light cracking due to use in boiling water. The spoon has many good years of life left. Small cracks in wood cookware are generally not concerning. *Image courtesy Letty Camire*

Poplar spoon finished with walnut oil. *Image courtesy Kristen Pickens*

Olive wood spoon and spatula, in service for one year, showing normal wear that is hidden more than on beech due to the natural color variation in olive wood. *Image courtesy Letty Camire*

## BUTCHER BLOCK COUNTERTOPS

A butcher block countertop is wood installed as a countertop, generally not heavily finished, with the purpose of being a sort of permanent cutting board in the home. They are beautiful—especially early on—but are also a curse.

This use of wood is tricky. It's tricky because inherently one side of the wood is covered, so air movement can't be equal. It's tricky because there is already a lot of glue in the wood, and it's tricky because, unlike a cutting board, if you screw up care, it isn't easy to fix.

Butcher block countertops do have the potential to work fine without conditioning if you want to take full advantage of the wood, but that means being careful with spills and other cleaning (never oversaturating the wood with water as you wipe it down). Since most of these countertops are glue-ups, this is an instance where mineral oil as a conditioner makes sense. Because the oil takes up sorption sites and void space, it helps protect the wood against spills and thereby also against dimensional changes. It may or may not impair the movement of bacteria down into the wood, which may then necessitate other cleaning measures other than just water wipe down.

There's no easy answer for butcher block countertops, and not a really right way to use them. Their use will always be a give and take, unless the homeowner can afford solid wood slabs—which come with their own issues in terms of warp. What is perhaps not as advisable is conditioning the countertop with a hardening oil, or an oil wax emulsion that builds a film. If the desire is to completely seal off the wood, then the countertop isn't a *butcher block* countertop, it is a regular countertop, and should just be coated in shellac or lacquer and be used as such.

## WOOD BOWLS, CUPS, AND CUTLERY

Many times wood used as table service has a history and has been passed down as family heirlooms. It is becoming slightly more common to see parents purchasing wood sets for their toddler children, but generally wood does not serve a major role on the table. How it is cared for and how it is finished, of course, depend on its use.

As with all wood in food contact, it is important to ensure that your wood is food safe before use. Outside of this important rule, how this type of wood is cared for is a balancing act between aesthetics, function, and cleanliness.

For *dry food*, such as popcorn and crackers, there is little need for a finish, although having one is unlikely to matter, since there is generally little concern over the bacteria on a cracker. Leaving the wood unfinished makes for easier cleaning if the goal is fewer microbes on the surface, but for dry food, a simple dusting with the hand is enough to knock off crumbs and restore the piece.

For *wet food*, such as some cheeses, fruits, etc., a finish isn't necessary and may in fact exacerbate discoloration. Cutting things such as strawberries can leave red stains in the wood, and in an unfinished cutting board, those stains usually disappear after another few wipe downs (as the coloring is drawn deeper into the wood). Unless the finish on the piece is thick enough not to allow penetration (and at that point, you might consider just using plastic), the colorants can often become trapped in the finish layers and remain.

For *liquids* a finish almost always must be used, especially if the liquid is an alcohol. Wood cups in particular—since the liquid will likely sit in them over time—need to be finished enough that the vessels are almost entirely filled. This, of course,

Handmade wooden bowl, unfinished, and used for three years as a popcorn bowl. The oils from fingers and the popcorn can be seen on the sides. Most food oils are nonhardening oils, and the fingerprints will fade with washing. *Image courtesy Letty Camire*

will directly affect the ability of the wood to move water (as in it no longer will), so cleaning will change.

Bowls for soup, especially hot soup, should be carefully considered. Many film-based finishes do not do well with heat and may break down over time (and in your soup). Oil finishes may be the best solution, but these take a very long time to cure.

For any use, if the vision is shiny wood, either a polish can be used—such as a wax—or the wood can be sanded to a very high grit (1200 or so). Getting the wood wet is likely to raise the grain and gum the wax over time, so these are temporary solutions. For wood that needs to stay absolutely shiny at all times, a high-gloss finish would need to be used. These tend to be film-forming finishes (see chapter 8 for more on wood finishing) and will almost certainly impair water movement.

Wood makes fantastic rolling pins. Same rules apply: don't finish unless you want to potentially lose those antimicrobial properties. *From the collection of Seri Robinson*

Unfinished curly Oregon maple cutting boards, cut from a board bought randomly from a box store. Cutting boards don't have to be fancy and expensive. The board here cost $10, and seven cutting boards were made from it in a little under an hour. *From the collection of Seri Robinson*

## BAMBOO

Bamboo is a *monocot*; not a *dicot*, such as wood. It is *woody* but doesn't have the same type of cellular arrangements. Because of how it grows, almost all bamboo in the kitchen is a glue-up of slim strips. With a significantly different anatomical structure to wood *and* copious glue, generally bamboo does not react the same way to microbes. Some species of bamboo are reported to have antimicrobial properties, but a discussion of such is beyond the scope of this book.

## WHEN TO REPLACE

Nothing lasts forever. At some point, wood kitchenware will need replacement. On the plus side, it is biodegradable and compostable. Like wood decks, consumers often replace their wood well before its service life is over. Images in this and previous chapters have shown wood in various stages of wear. Of all the photos shown, only the rubberwood cutting board that split its glue line actually needs to be replaced. Minor cracking in wooden cutting boards and spoons is of little concern for continued cooking.

1. N. O. Ak, D. O. Cliver, and C. W. Kaspar, "Cutting Boards of Plastic and Wood Contaminated Experimentally with Bacteria," *Journal of Food Protection* 57, no. 1 (1994): 16-22.
2. D. O. Cliver, "Cutting Boards in *Salmonella* Cross-Contamination," *Journal of AOAC International* 89, no. 2 (2006): 538-42.

# Chapter 4
## WOODEN TOYS

Toys! It's so easy to accumulate bins and bins of forgettable plastic, but there's something about wooden toys that really sparks for parents. Whether its our own fond memories of wooden toys, or something more primal in the material, wooden toys have remained an enduring (if not often expensive) gift for children.

With all the worry about the leaching nature of plastics, wood can seem like a safer alternative for teething. After all, it's natural! As has already been discussed, wood has plenty of extractives that can leach out, especially given a hot, wet environment such as a mouth. This chapter deals entirely with wood from a toy perspective, in terms of what is likely safe, what isn't, and how to source safe, high-quality wooden toys.

Young child on a Pikler triangle made of European pine (Finland) and some Baltic birch plywood. Pikler triangles are a common wooden toy used to help gross motor skill development. This image is from the Philippines, where toys like this are rare. To help with their daughter's development (Zen, the little girl pictured in the photo), the family made their own Pikler triangle. *Photo courtesy Brian A. Tan of Developmental Depot*

**Young child with a cherry teething toy.** *Photo by Cecillia Otake; image courtesy Sarah Scott: www.pacicatchers.com*

## TOYS FOR TEETHING CHILDREN

Generally, any wood that is suitable for cooking or kitchen use is also suitable for teething children. Some extra caution should be observed for woods such as black walnut, where exposure to toxic extractives would likely not be problematic for an adult but may have the potential for an adverse reaction in a much-smaller body.

As with cookware, no finish is the best finish. Think of wooden toys like cutting boards and your kid's mouth like raw meat. Nothing beats a perpetually clean surface, and with raw wood, you get just that (given enough set time, of course).

Most wooden toys do come with a finish, which just means their antimicrobial nature can't really be taken advantage of. Many Etsy sellers are perfectly happy to leave the finish off (saves them time and money!), so if you're shopping handmade, go ahead and ask in advance. Worst they can say is no.

**Girl with a wooden camera.** *Photo by Just the Jams Photography; image courtesy Sarah Scott: www.pacicatchers.com*

## TOYS FOR HAND PLAY

Almost all wood is safe for hand play. There are some woods that can cause contact dermatitis, and some woods with unsafe smells (such as cedar; see appendix I), but these are in the minority. Buying a wooden toy for hand play should hold very little concern for the purchaser, especially if the wood is finished. With a finish (*not* a polish), even troublesome aromatic extractives can be contained.

## CLEANING WOODEN TOYS

Again, kitchen rules apply. If the wood is unfinished, a wipe with a wet cloth or soapy wet cloth and then setting it to dry for a few days should work fine. If the toy has a finish on it, use a soapy cloth and a gentle rinse and then let dry for several days. No dishwasher. Ever. Bleach, vinegar, and other such things could potentially damage the finish, so these are also not ideal. Soap does a great job of flushing off microbes, but if the whole prospect of microbes bothers you, unfinished wood might be your best bet.

## WHEN TO REPLACE WOODEN TOYS

It's rare to need to replace a wooden toy, outside of the toy being broken. Most well-made wooden toys should be fine to hand down as generational heirlooms. Consider replacing wooden toys that are broken to the point where simple wood glue can't fix them, or where the finish is chipping and you are unaware of the toy's origin.

A light dusting of mildew or mold on the outside of a wooden toy is generally *not* a cause for concern. Gently scrub the toy with soapy water, rinse, and let dry. Some discoloration may remain, but as long as the toy stays dry, the fungi cannot continue to grow.

## TOYS FOR INDOORS VERSUS TOYS FOR OUTDOORS

Wooden toys, in theory, are fine for indoor and outdoor play. If you live in a climate where the interior relative humidity and temperature are controlled (heat/AC), then taking *unfinished* wooden toys outdoors may cause some problems with shrinkage and swelling. For toys you plan on having many such transitions, it may be best to stick with commercial wooden toys that have heavier finishes. Alternatively, a high extractive wood that cannot take on as much water would work well (such as teak).

There are some woods, such as cedar, that make for great outdoor use but not indoor use, due to their aromatic extractives. Cedar is fantastic for outdoor toys due to its natural decay resistance. Outside, its smell is less of a problem (wind), which is why play structures are often made from cedar. If you're considering sandbox toys that will permanently live outside, or other such toys, naturally durable woods make for an excellent choice.

Child using a teething ring outdoors. Because teething rings are meant for wet and dry conditions, taking one outdoors should not present a problem. *Photo by Bree Larsen; image courtesy Sarah Scott: www.pacicatchers.com*

## WOODEN TOYS FOR WATER PLAY

Wooden toys can certainly be used for the bath or water tables / pools, provided that they are dried properly afterward. For these types of uses, *unfinished* wood often works better since it can dry faster. Finishes tend to slow the absorption of water but also slow its evaporation, which leads to molding. Unless your wooden toy comes with a marine-grade varnish on it, it's best to use unfinished wood for water applications and then let dry, propped, the same as you would a cutting board.

## ANTIQUE WOODEN TOYS

Because wooden toys are so durable, many antiques are still in the hands of children. With unfinished wood and a lot of the basic finished wood, this is little concern. Unfinished wood remains antimicrobial, and even older finishes are usually nontoxic once cured.

Where you will want to be careful is with older paints, since historically many contained lead. Old wooden toys with paint should be tested with a lead kit (available at many box store merchants) to ensure safety before even hand play is allowed.

Wooden fishing game made from several different wood species. While some wooden toys may look like they were designed for water play, check with the manufacturer before using them in water. *Image courtesy Sarah Scott: www.pacicatchers.com*

## HOW TO FIND WOODEN TOYS

Appendix II offers a list of manufacturers of wooden toys—both small and large—who are known both to (1) sell wooden toys and (2) think about the application of the toy in their wood selection. In general, wooden toys are once again gaining popularity and can be found from major retailers to places such as Etsy. Adding the modifier "wood" or "wooden" to your search term (so instead of baby gym, type "wooden baby gym") will pull up any number of brands. The results on Etsy are endless.

What you'll want to watch for are products made from composite materials—such as plywood—versus products made from solid wood. Composite materials may have components that are not safe for teething, so when shopping for a teething child, it's best to err on the side of solid wood. For hand play there is little concern, so composites can be a nice way to save money and weight on larger play structures.

Rainbow rockers are another common wooden toy in the US. When on end they can be used as climbing bridges. When on their backs they can be used as rockers. Most are made from a combination of plywood and solid wood. *Rocker made by www.etsy.com/shop/sensoryplay, from the collection of Seri Robinson*

Other common wooden toys are stacking towers (giant stacking tower pictured manufactured by Grimm Toys, www.grimms.eu) and rainbow stackers (pictured as taken apart, made from rainbow poplar and unfinished by SensoryPlay: www.etsy.com/shop/sensoryplay). *Both toys from the collection of Seri Robinson*

Wooden peg dolls, often painted and finished, make excellent toys for hand play. *Peggies made by https://tiffanyleestudios.com, from the collection of Seri Robinson*

One of the unique qualities of using wood for toys is that the maker can mimic larger structures. This boat toy chest showcases flawless attention to detail while still being functional in a play space. Made by Adamzoriginals: www.etsy.com/shop/Adamzoriginals. *From the personal collection of Seri Robinson*

While finishes and paints can help wooden toys compete with plastic, a lot can be done with just simple woodburning (pyrography). This Forest Friends play set contains a wooden tube (spar jointed) with one end open for storage and the other end covered with hide to make a drum. Slots were cut in the bin to house the forest creatures, which are simple bandsaw animals with woodburning. Forest Friends was a group project made by students at Oregon State University. *From the collection of Seri Robinson*

A blend of paint and raw wood is common in many commercial wooden toys that must compete for shelf space with plastic. Pictured are two stacking bears by Haba (www.habausa.com). *From the collection of Seri Robinson*

Wooden cars and trucks are a classic toy, shown here with natural lacquer wheels. *Toys made by http://littlelatitude.com, from the collection of Seri Robinson*

# Chapter 5

## CLEANING WOOD IN THE HOME

Wood is surprisingly easy to maintain in the home—much more so than cleaning-product companies would have you believe. The first hurdle is to realize that wood does not naturally shine unless sanded to a very high grit in sandpaper (over 1000), or finished with a high-gloss finish.

Much of the wood in service in the home is cleaned due to looking "dull," which many mistake for being dirty. There are a number of reasons why a once-shiny floor or cabinet has lost its gleam, and almost none of them have to do with dirt. Below are the most commonly complained-about areas in the home, and some thoughts on how best to clean them.

Home setting with multiple uses of wood, from the bookshelf to the stand to the floor. The bookshelf was darkened from its natural color by using a vinegar, steel wool, and tea combination. A water-based polyurethane was used for the finish. *Image courtesy Candace Hermann: https://stayingincourage.com*

A 1930s antique cabinet from Hong Kong made partially from teak. With multiple layers of stain and finish, and deep carvings, this piece requires some protection with every cleaning. A general dusting works week to week, but a cleaner with a shiner would be worth using once every other month as long as it didn't get into the carved areas, where it could easily build up. *Dresser from the collection of Tai Labrie*

Detail of the drawer pulls of the Hong Kong dresser. While cleaning, care should be taken to avoid getting polish inside the reliefs. *Dresser from the collection of Tai Labrie*

## DUSTING AND POLISHING FURNITURE

Generally speaking, furniture seldom, if ever, needs to be cleaned. Wooden furniture does need to be dusted regularly, but a microfiber dust cloth can do this quite well. For older furniture with burrs still on the wood, a feather duster will work. Try to avoid spray polishes unless the wood really gets a lot of use, since these almost uniformly have waxes or other types of shiners in them that build over time. This buildup can eventually dull the finish of the wood, or even form a very apparent plastic-like layer that is difficult to remove.

Side carving on the Hong Kong dresser. Due to the size of the carving, some cleaning needs to be done internally, but since the carving should see little wear, only a cloth and a bit of water should be used if needed. *Dresser from the collection of Tai Labrie*

A captain's chair from a secondhand store in Rhode Island, US. Older style, clearly used, but with minimal wear. *Image courtesy Letty Camire*

Should a layer such as this form, or you inherit a piece that already has such a layer, discontinue use of the polish. Some polishes can be worn down over time (they aren't generally superhard layers). Some you might be able to strip off yourself, using alcohol or vinegar—but be warned, since this could have unintended consequences if your wood has a stain on it. The best course of action, especially if the wood is antique, is to take it to a professional furniture restorer who can treat it with the care it needs. If you do go this route and you like your wood shiny, ask the restorer to put on a high-gloss finish for you, and boom! No polish ever needed again.

The handle of the captain's chair shows some wear to the finish and the stain. A cleaner with a shiner would be appropriate to use in high-wear areas such as here, and on the seat. *Image courtesy Letty Camire*

A three-month-old white oak floor. The floor was improperly installed and did not account for changes in relative humidity, and now it creaks and pops when walked upon. Care must be taken to minimize water on wood floors, especially those already under stress from improper installation. *Image courtesy Cassandra Winters*

## MOPPING FLOORS

Wood floors seldom, if ever, need to be mopped, and they definitely do not respond well to frequent mopping. Remember, wood swells when it gets wet, and no finish is 100 percent waterproof. Add to that the increasing number of wood floor cleaners that utilize steam to clean (this is a *terrible* idea), and you can, I hope, intuit why you should stay away from such methods.

Most wood floors benefit from a "no shoes" rule (keeps mud and such from being tracked in), as well as a spot-clean rule. There are any number of microfiber pads one can buy these days that get the floor just lightly damp, and these, along with things such as baby wipes, are ideal for a quick spot clean for a spill.

Resist the urge to "sanitize" your floors. There's no way to truly do this with the products currently on the market, and, really, a few licks to a floor never killed anyone (I hope). Most attempts at sanitation end up with ruined finishes and a need to completely refinish the wood. It's just not worth the cost.

There are a number of products for cleaning wood floors, just as there are for polishing furniture, that are not wise to use. Many of the wood floor-cleaning solutions on the market contain shiners that will build up over time if your floor is not a high-traffic floor. Some manufacturers market this on purpose (anything that promises to restore your floor to a healthy shine, for instance), and these should *never* be used outside of perhaps preparing a home for sale, where you will mop once just to get the home looking its best.

Stained oak laminate flooring with one application of a polishing wood cleaner. The floor had never been mopped prior to this application. The film layer from the cleaner is clearly visible from where the mop didn't quite meet the carpet.

When shopping for a wood floor cleaner, pull the SDS (safety data sheet) from online or from the retailer. All are required by law to have them available. Check out the percentage of alcohol (higher tends to be better) and what cleaners are in it, as well as the percentage of other components (these are usually the waxes or other shiners).

Unfortunately, there isn't one cleaning solution that works best, since there are so many finishes on the market today and each responds a bit differently. For instance, older floors that have their original finish intact should be cleaned only with soap and water, since alcohol can dissolve the shellac. Newer floors that claim to be "sealed" often come with a specific shine level: matte for the "hand scraped" (it's not actually hand scraped; the boards are milled like any other and then go through a distressing machine to make them look that way), and high gloss for more "finished" floors. No amount of product will make a matte-finish floor look like a high-gloss floor. And it can be very easy to turn a high-gloss floor into a cloudy finish with a steam mop, where the steam gets under the finish and then fails to evaporate at the needed speed. The heat doesn't do good things to the finish itself, either. High-gloss finishes cloud easily when additional shiners are applied, meaning that using most commercial cleaners can result in a film buildup that slowly, over the course of several months or years, results in a dull, matted floor.

Removing this buildup can be done in a number of different ways. The obvious one is to hire a flooring refinisher, although this is cost prohibitive for many people. There are some cleaners on the market with a fairly high alcohol content that can strip the buildup over time, but again, this can take months. You can try to speed up the process by scrubbing the finish by hand with a high-alcohol cleaner, but be warned that this could also very well take off the original finish. For those who can't refinish, switching to a product without shiners is ideal, as is simply ceasing mopping. Those built-up films *will* degrade over time with use, and there's no reason you can't let it happen naturally.

## WOOD IN THE BATHROOM

As much as I love wood, it really doesn't belong in the bathroom unless finished with a marine-grade finish. A bathroom is a perpetually changing environment, going from normal indoor humidity and temperatures to hot and wet, sometimes multiple times a day. This is *very* hard on wood, especially wood that frames around something such as a sink, where liquid water would also be added to the equation. It can be very easy for wood in the bathroom to begin growing mold and mildew unless the space is very well ventilated.

Should growth begin to occur on your wood in the bathroom, wipe the offending microbes off with something disposable, such as a baby wipe, and *get the humidity down*. Fungi can't grow without moisture, and no amount of bleach or other sterilization methods will matter if you can't get the humidity under control. If replacing the wood in the bathroom is outside your budget, consider either refinishing the wood so that it takes on less water—such as with one of the thicker marine-grade varnishes—or put a dehumidifier in the space.

## WHAT TO DO ABOUT MOLD ON WOOD

DO NOT PANIC. There are a lot of wild urban legends flying around about mold, especially "black mold." Please remember that fungi are *an entire kingdom*, just as animals are a kingdom. The animal kingdom encompasses organisms from humans to fish to spiders, so consider how ridiculous it would be to say that you were once bitten by a spider so you want to kill all cats. One or two offending fungi that very rarely grow in any sort of concentration to do damage are no reason to call for a mold remediation team.

Many people do, of course, have *allergies* to some molds. Many people also have allergies to cats. As with cats, if you find that someone in your home has a hard time breathing in the bathroom, there is no need to panic. Get the humidity down and clean the offending area, and the fungi will eventually die. That's really it. If it makes you feel better to use stronger cleaning products, go for it, but they aren't doing anything above and beyond what simply removing the moisture would do. Be warned, too, that strong cleaning products can damage the finish on wood, so once again, these are best avoided unless absolutely necessary.

Bamboo cutting boards. The round one shows separation along the glue line due to too much water being applied. When cleaning any wooden object, care must be take to minimize water contact. *Image courtesy Cassandra Winters*

# Chapter 6

## DECKS, OUTDOOR PLAY STRUCTURES, AND PATIO FURNITURE

Wood and water interact, and even the best finishes can't keep the elements out forever. Knowing what to expect out of wood exposed to the elements can save a lot of money, since many people replace their outdoor wooden structures long before they need to—almost entirely on the basis of aesthetics. This chapter breaks down the most-common issues encountered with outdoor wood, what to look for, and when to replace.

## DECKS

People in the US *love* decks. The suburbs, in particular, are known for a deck in every backyard. There are dozens of materials available to build decks, but wood remains an enduring favorite.

Even within wood products a number of options are available. There are wood/plastic composite decks, made from a mixture of wood dust and plastic, that are thought to have a higher decay resistance than solid wood. There's pressure-treated wood, which is usually a pine of some form that has been treated with a copper chemical to give an otherwise low-decay-resistant wood some longevity. Then there are the naturally decay-resistant woods such as teak and cedar that are used directly, sometimes even without a finish.

The more naturally decay resistant the wood, the longer it will last. For decking especially, it's important to make sure that the wood in ground contact has the most resistance, whether that means submerging cement blocks and then putting support beams on the cement or using pressure-treated wood for the ground contact wood and building with cedar on top. A well-made deck, whether from decay-resistant wood or pressure-treated wood, should last *decades* before decay is severe enough to warrant replacing.

Unfortunately, many homeowners equate the graying of wood to decay. Wood *naturally grays in sunlight*. UV light breaks down the lignin in wood, causing the color change, and for the most part this process is in no way harmful to the wood structure. It can be unsightly, sure, and there are tons of products on the market meant to slow the graying down, or to cover it up with various pigments. All work to some degree, but eventually your deck will turn from its native wood shade to gray. And much like with a dull shine on a wood floor, *this is okay*. Graying is a natural part of wood's life cycle. Do not panic, and certainly don't let someone talk you into replacing your deck. Until you see visible signs of rot in the supporting members or anywhere near ground contact, your deck is likely fine.

In that same vein, if your deck becomes mildew covered or sticky, resist the urge to power wash. Power washing can damage the wood, selectively removing the earlywood (the less dense part of the growth ring) and leaving the latewood. This results in an uneven surface on your deck and can make the wood look far more weathered than it actually is. Instead, consider just using a garden hose to spray the deck down, or spot cleaning, or waiting for a good, hard rain.

Outdoor wooden playhouse with finish peeling. There is no structural damage to the house despite the damaged finish. The graying is from UV exposure and does not pose a structural risk.

Unfinished cedar boards used to make a permanent outdoor sandbox. After five years in service there is still no decay to the boards.

## OUTDOOR PLAY STRUCTURES

Because these are meant for children to use, be a bit more liberal with your inspection of the wood. Make sure that the wood in ground contact is firmly sound and that none of the joints are slipping. The cedar play sets on the market today, in particular, are quite durable, but that doesn't mean that several grade-schoolers scrambling over the structure day after day won't cause some wear. Watch for bowing as well, since play structures, like bookshelves, experience a very specific, repetitive load and, with the moisture cycling of the outdoors, can experience hydrogen bond slippage. It's not a major concern, but over the course of several decades could lead to failure of the member.

## PATIO FURNITURE

You'll be hard pressed to find patio furniture that doesn't have some kind of finish already on it unless you're shopping in the more expensive exotic woods, such as teak. Don't expect the factory finish to last more than a year if you leave the furniture outside, but know that when the finish begins to peel, that doesn't mean the set has rotted. Even non-decay-resistant species have a good half decade in them, if not longer, if setting on a concrete slab. If you'd like a bit longer life, sand down the old finish and apply a new one once every summer, and keep the wood protected. Otherwise, let it peel and go gray, and when you see the bottoms of the feet start to rot, toss (or burn) the set and get a new one.

One of the best features of wood, after all, is how renewable and biodegradable it is!

Eucalypt patio furniture with three years of constant outdoor exposure in the Pacific Northwest of the United States. The finish is peeling, but the wood remains sound.

# Chapter 7

## QUICK AND DIRTY WOOD REPAIR

The following is a pictorial guide to some common wood issues, along with tips for repair.

**The fundamental tools of quick wood repair: drill, painter's tape, tape measure, hammer, dowels, and glue.** *Photo courtesy Candace Herrman: https://stayingcourage.com*

## CUTTING BOARDS

Glue-up cutting boards that have delaminated and warped are often more hassle to fix than they are worth. If your board is a family heirloom, your best option is to finish separating the panels (either by pulling or cutting apart on a bandsaw), re-edging them on a joiner, planing the boards flat, and gluing the whole thing back up. For boards too small to go through a planer, you can attempt to bend them back to shape by separating them, cleaning off the glue, and putting them back in the dishwasher. Immediately upon the cycle end, remove the boards and clamp them to a flat surface, forcing them to dry in a somewhat flatter arrangement. *Photo courtesy Carol Johnson*

Cutting boards with deep score marks on the surface can be sanded down with a random-orbit sander and some 150 paper to reflatten the surface. For mold residue on the edge from improper drying, hand sanding with some 150- or 120-grit paper should remove most of the discoloration. Neither the scoring nor the fungal pigment generally represents a cause for concern, and their removal should be done primarily for desired aesthetics, if at all. *Image courtesy Letty Camire*

General discoloration on cutting boards isn't something to worry about. For aesthetics, the board can be uniformly oiled with a nonhardening oil—such as mineral or a vegetable oil—to even the color, but be aware this may dampen the wood's antimicrobial properties. *Image courtesy Meredith Spitalnik*

Glue-up wooden cutting boards that have started to delaminate but haven't finished the process are very irritating. You can attempt to repair the glue line by buying some white wood glue and running it into the cracks, then clamping the board together overnight. Generally, it's best to just chuck the board and buy a new one that isn't a glue-up. *Image courtesy Meredith Spitalnik*

## TABLETOPS

Dents in soft woods, especially in tabletops, present no real harm to the furniture but are unsightly. The best option in these scenarios is to sand the tabletop down with a random-orbit sander and some 180 or 220 paper (so, a gentle sand), then refinish with a topcoat finish that will lend a bit more durability to the wood surface. *Photo courtesy Carol Johnson*

Marks from pens on harder woods can often be removed with a magic eraser or a bit of floor cleaner with a higher alcohol content. On softer woods, such as the table pictured, the pen has dented and left color, leaving no option but to sand down. *Photo courtesy Carol Johnson*

Toward the back of this photo you can see cloudy patches from where a hot item was placed in contact with the finish. Sometimes the cloudiness can be scraped off with a fingernail, but if the item was sufficiently hot, the damage is all the way through the finish. Like with the cutting board on p 69, the best option is to sand down. No finish can 100% protect against heat, but oil finishes are less likely to film like this than water-based ones. *Photo courtesy Carol Johnson*

The finish damage on this table was caused by a tomato rotting over a cloth that sat on the table for several days. The moisture from the tomato eventually got under the finish and clouded it (as did some of the breakdown reactions from the rot). As with the other image (*above*), the best option is to sand down and refinish. *Image courtesy Meredith Spitalnik*

Small, shallow scratches on dark-stained wood can't be repaired without sanding but can easily be hidden by an application of oil (e.g., walnut, tung). Don't use oil with a solvent carrier, or you risk damaging the current finish. *Image courtesy V. Thiem*

Sharpie marker on a maple tabletop. If gotten to right after application, the mark can usually be rubbed off with a magic eraser. If left to set, especially on worn finish, stripping the finish off will be required. If on raw wood, scrubbing with a wet rag can help move the color into the wood, eventually hiding the mark. *Image courtesy Candi King*

# FURNITURE

On antique furniture where the joints are separating, little can be done without taking the structure apart. Dovetail joints (pictured) are surprisingly strong and can hold against warp well. Reinforcement via nails/screws (also pictured) isn't necessary, generally, but doesn't hurt anything if it makes you feel better. Older pieces, especially those from Japan, may not have been glued (because the joints were expected to hold on their own), so if you get a higher-end piece and it warps, in theory you could knock the pieces apart gently with a mallet, straighten the wood with some steam and clamps, and put it back together again. I'd advise against this and suggest just taking the piece to a furniture restorer instead.

Thin backing peeling from an old sewing desk. The damage was likely caused due to water exposure that swelled the wood and caused the warp. Some discoloration has formed from where the stain leeched, and there is some mold on the end grain of the adjoining piece. Damage like this, while likely not structural, should be dealt with by a carpenter or furniture restorer. Cosmetic fixes would include sanding off the fungi and sanding down the back sheet (past the stain), then restaining and refinishing. *Image courtesy Jess Laxo*

Peeling veneer from a tabletop. This damage is bordering on too extensive for home repair. You could try gluing and clamping to stick it down, but likely the entire sheet will need to be removed and a new one placed by a professional. *Image courtesy V. Thiem*

## OUTDOOR STRUCTURES

A rotting dock. This shouldn't be repaired—the components need to be replaced, but likely the posts (in the water) are fine, assuming they were treated properly or were made from a durable wood such as teak. *Photo courtesy Donald McClure*

Outdoor wooden structures naturally gray with UV light exposure. This isn't something that needs repaired, but if it bothers you, you can always paint it. *Image courtesy Sarah Jensen*

This image shows not only the graying of the wood, but also the peeling of the finish. Before applying a new finish, make sure the old one is entirely off by sanding the wood down. Resist the urge to use a pressure washer. The spray can remove wood as well as finish and, on damaged wood, can take off some of the earlywood, speeding the degradation of the structure. *Image courtesy Sarah Jensen*

Rot in ground contact is very common and should be taken seriously and addressed by a professional, especially on load-bearing structures. *Image courtesy V. Thiem*

Even on non-load-bearing areas, rot on ground-contact wood should be taken seriously, since fungi can easily run up the member and onto structural areas. Seek professional guidance in these cases. *Image courtesy V. Thiem*

Outdoor wood that was bleached (bleach was applied) after repeated dog urination (the lighter area). Wood isn't damaged, but the bleach did override some of the graying. Repeated bleaching would damage the wood. *Image courtesy V. Thiem*

This exterior structure is starting to show more than just general climate-fluctuation cracking. It should be inspected and potentially replaced, not repaired at home. *Image courtesy V. Thiem*

# FLOORS

Typical wear to an oak floor. Over time, the softer areas of wood wear down more quickly than the harder areas (often, with earlywood wearing down before latewood). Here the damage to the finish and wood was caused by a dog's toenails. The easiest way to fix is by sanding down and refinishing. *Image courtesy Sarah Jensen*

Another example of the uneven wearing of a wood floor. The less dense earlywood (less dense because it has thinner cell walls usually) wears before the latewood, resulting in this uneven pattern. Sanding down and refinishing is the only option for repair. *Image courtesy Candi King*

# MISCELLANEOUS

Broken bowls that separate cleanly, such as in this image, can simply be glued back together by using a combination of belt clamps and bar clamps. *Image courtesy Letty Camire*

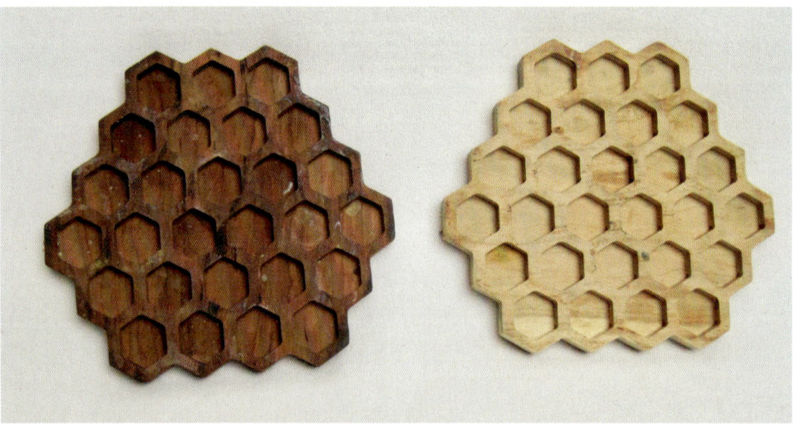

These trivets were made at the same time, of the same cherry wood. One sits on the countertop all day and is in constant use, while the other sits in a drawer. The darkness of the more used one is due to a combination of heat and daylight (UV) exposure, both of which naturally darken wood over time. While this is not really "damage," you can remove the darker color with a simple light sanding. *Image courtesy Letty Camire*

Spoon bent by dishwashing. Run it through the dishwasher again and straighten by hand as it dries, or clamp flat right after it comes out. Be careful not to be too forceful or you will snap the wood. *Image courtesy V. Thiem*

By itself, finish peeling is easy to fix—simply apply a new coat of finish. In this case, a topcoat finish that yellows would be appropriate, such as a lacquer. However, the exposed wood has grayed due to UV damage, so finish over it should not be done until the gray is removed. In this case, the sill should be sanded with 100-grit paper to remove the finish and the UV damage, then refinished. Since the sill is near a sink, an oil-based finish would be more appropriate, or an oil-based finish followed by a water-based topcoat. A marine-grade finish would also be appropriate. *Image courtesy Meredith Spitalnik*

Hand sanding with some 120 paper could ease the edges of this chip, but it will never be completely removed. If the knife cannot be replaced, the divot could be gently opened with a Dremel grinder, and a tiny piece of wood inlaid (and glued in) over the hole. *Image courtesy Sarah Jensen*

Cracks through structural supports should always be replaced by a professional. *Image courtesy V. Thiem*

Painted, finished toys with simple breaks can be glued and clamped. Resist the urge to use less glue for fear of getting it on the paint. A starved glue joint will just crack again. You can always scrape the excess glue off after it has set, and repaint.

# Chapter 8
## FINISHING WOOD

Wood finishes can be broken down into three categories: surface, penetrative, and polish. A surface finish builds a film on the surface of the wood and only marginally penetrates the cells. Shellacs, some lacquers, and water-based polycrylics fall under this category.

Penetrative finishes are finishes that penetrate deeply into the wood and take many coats to build even a semblance of a film. Most penetrative finishes are urethanes or oils (or a combination of the two). It's important to note that within the oils there are two types: hardening and nonhardening oils. Hardening oils will cure over time, forming a (semi)permanent film inside the wood. Nonhardening oils will remain liquid and may move through the wood over time. Nonhardening oils do not, generally, form any sort of protective coating, but they do fill up binding sites and cell void space, preventing the wood from taking up as much water.

Polishes are usually made from wax and form a nondurable film coating on wood, with little to no penetration (oil-wax emulsions penetrate a bit). Outside of carnauba wax, which is difficult to apply without a buffing wheel, waxes mostly make the wood look shiny and impart little, if any, water resistance to the wood for more than a few uses.

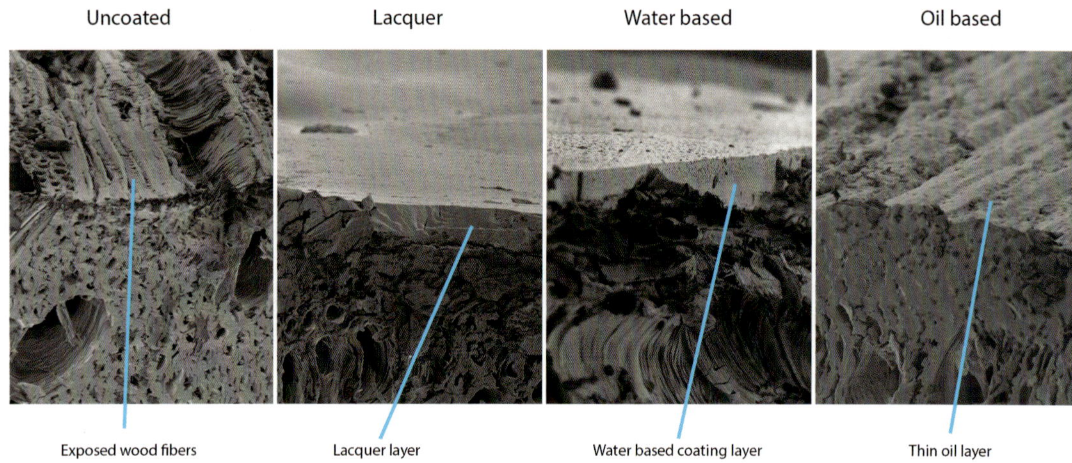


SEM images of coated sugar maple (*Acer saccharum*). The uncoated sample shows the raw wood face. The lacquer-coated sample shows a thick cover with just a bit of penetration into the wood fibers. The water-based sample shows the finish sitting almost entirely on top of the wood in a thick layer. The oil-coated sample shows a very thin cover layer with deep penetration into the fibers. *Image courtesy Sarath M. Vega Gutierrez*

There is some overlap between categories. Some lacquers are so heavily solubilized that they function as a sort of penetrative topcoat. Many toymakers mix oil and wax to form an emulsion that gives toys a bit of deeper protection than the wax alone.

When determining which finish to use for a given wooden object, it is critical to consider the object's use. Will the wood be outside and constantly exposed to the elements? Will it be used in a bath and have ample time to dry? Will it be in food contact? Understanding how you want the wood to perform *first* will help in deciding what finish to use.

## WOOD FLOORS

This one is more personal choice than anything else. Some people choose to leave their wood floors unfinished and court the dangers of spilled liquids and seasonal swell. Others prefer thick surface finishes that attempt to "seal" the wood from moisture. Others like the shine of a thick oil coat with a wax on top. Know that whatever you choose, wear and upkeep will be critical. Waxes are generally not durable but are easy to reapply and set quickly. Penetrative finishes give a long luster, but if something messes up the finish, generally the entire slat has to be replaced. Surface finishes can flake and crack but can be repaired through sanding.

A new option on the market for finishing wood floors is prefinished flooring. This is where the finish is installed at the factory and cured ahead of time. These finishes have the benefit of being very hard and stable and are done off-gassing by the time the floor is installed in your home. If they do get scuffed, they are hard to repair by a general flooring contractor, but they are also harder to scuff. If you don't want to deal with the weeks of smell from a traditionally finished wood floor, consider prefinished flooring.

A handy flowchart is included for reference.

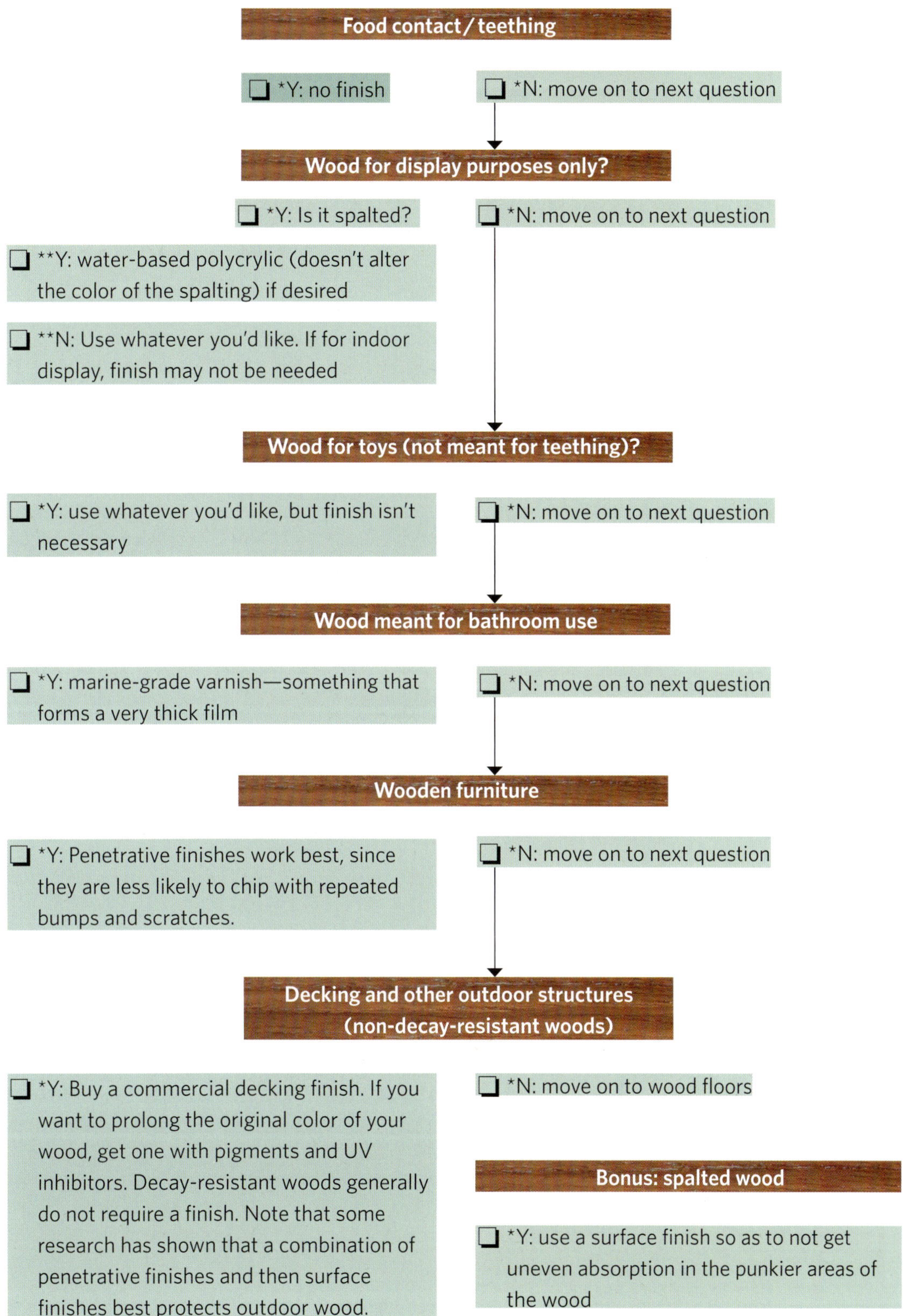

# Chapter 9
## COLORING WOOD

There are several common ways to color wood, and which you use depends very much on the purpose of the object. Many general wood colorants are not rated for food safety and, as with any additive, will take away from the antimicrobial nature of wood. As such, consider the purpose of the wood object before deciding to color, especially with wood that is meant for toys or food contact.

Painted poplar toy by Kristen Pickens of Two Raccoon Hollow

## STAINS

Stains, generally speaking, are commercially produced compounds meant to darken wood. They come in and out of fashion but have been used for hundreds if not thousands of years to change the appearance of wood, especially for wooden furniture. Most stains turn the wood darker brown, sometimes into shades of red or black.

There are a few DIY and natural stains you can use, such as using tea bags or coffee grounds, running a piece of steel wool across a wet piece of oak, etc. Be aware that using food items to color wood can introduce sugars or other components that can make the wood a more attractive breeding ground for microorganisms, and that many of these types of stains are not very deep, or very dark. On something such as a cutting board, a few washes will readily move a tea stain down into the wood, returning the board to its natural color. As such, natural stains don't tend to work well as permanent colorants unless a heavy-duty finish is used on top.

Pine boards before and after staining. Many things can be used to stain wood, from commercial stains to tea bags, and everything in between. *Images courtesy Candace Hermann: https://stayingincourage.com*

## DYES

Woodworkers have a long history of using aniline dyes, which can be purchased either as water-soluble or alcohol-soluble powders. Aniline dyes penetrate deeply into wood and give vibrant hues, but the colorants are not light stable and fade over about a decade, even in indoor light. It is always a good idea to put a finish over aniline-dyed wood, since the dye does not readily bind and can easily come off on wet fingers.

As with stains, some natural dyes exist on the market. These come mostly in shades of brown and black—occasionally red—and have similar issues to natural stains.

## PAINT

It is increasingly popular to see wooden toys painted with acrylic paint—especially watered-down acrylic paint, so that the color looks more like a dye. There are acrylic paints on the market that are rated for children, although note that care must always be taken with any compound that could potentially be brought into the mouth. Watered-down acrylic paint gives vibrant color to wood and is marginally more light stable than aniline dyes—and at this time has low to no safety implications for woodworkers.

Magic wands made from wood and synthetic dye. Many wooden toys are colored with watered-down acrylic paint, while many wood turnings are colored with water- or alcohol-soluble aniline dyes. *Image courtesy Amy Forman: www.etsy.com/shop/ChaosandRainbows*

Full range of colors possible from spalting fungi currently available. Image showcases the pigments from *Chlorociboria aeruginascens, Chlorociboria aeruginosa, Scytalidium cuboideum, and Scytalidium ganodermophthorum. Pigments generated at Oregon State University, Department of Wood Science & Engineering; image courtesy Patricia Vega Gutierrez*

## PIGMENTS

Pigments are rarely used to color wood, with the exception of spalting pigments. These colorants, derived from a specific class of wood-decaying fungi, have been used to color wood since at least the 1300s.[1] The pigments have long-term light stability (over the course of hundreds of years), are nontoxic, and are easy to apply (but not with water, since they are not water soluble). They are not cheap and are available only from a handful of outlets due to the difficulty in making them. At this time there is only one book in publication on the history and use of these pigments—*Spalted Wood: The History, Science, and Art of a Unique Material* (2016)—so those wishing for more information on the topic should consult that resource.

Oak (*Quercus garryana*) showing pink and blue stain from the fungus *Scytalidium cuboideum. Spalted by induction at Oregon State University, Department of Wood Science & Engineering*

Wild spalted blue green caused by the fungus *Chlorociboria* spp. (elf's cup). Found in Sitka, Alaska. The genus is widespread throughout the world, and most species produce this striking color. North America has only two known species: *C. aeruginascens and C. aeruginosa.*

Oak (*Quercus garryana*) showing orange zone lines and blue stain. *Spalted by induction at Oregon State University, Department of Wood Science & Engineering; turned by Seri Robinson*

Elm (*Ulmus* sp.) spalted in the wild and by extracted fungal pigments. *Turned by Mark Lindquist of Lindquist Studios and Seri Robinson of Oregon State University, Department of Wood Science & Engineering; photography by Mark Lindquist of Lindquist Studios*

Green zone line from the Amazon rainforest of Peru. Many zone lines appear black due to heavy color concentration but are actually made up of lower-weight pigments.

Red-purple pigment from the Amazon rainforest of Peru

Red zone lines from the Amazon rainforest of Peru

Cabinet from the National Museum of Decorative Arts, Spain (1600). Wide view. All blue green is from *Chlorociboria* spp.

Detail from one of the panels. National Museum of Decorative Arts, Spain (1600). The bird and leaves are blue green from *Chlorociboria* spp.

Close-up detail from one of the panels. National Museum of Decorative Arts, Spain (1600). The bird and leaves are blue green from *Chlorociboria* spp.

Close-up detail from one of the panels. National Museum of Decorative Arts, Spain (1600). The bird and leaves are blue green from *Chlorociboria* spp.

## NATURAL WOOD VARIABILITY

As an end note, wood comes in just about every color you can imagine. If you're willing to hunt for it, you can find purples (purpleheart), reds (bloodwood), yellows (canary wood), blacks (ebony), orange (yew, osage orange), and many other colors simply from heartwood. Take care, since these colors more often than not come from extractives and may be harmful.

1. S. C. Robinson, H. Michaelsen, and J. C. Robinson, *Spalted Wood: The History, Science, and Art of a Unique Material* (Atglen, PA: Schiffer, 2016).

Wood urn. Base made from spalted maple burl *(Acer macrophyllum)*, showing a typical beige wood color. Top made from curly walnut (*Juglans nigra*), showing the deep-brown color that walnut heartwood is known for.

# Appendix I

## "UNSAFE" WOODS FOR TEETHING

*Note:* This chart is not meant to be medical advice. It is not exhaustive and is meant only as a guide to further study. If a wood species does not appear on this chart, that does not make it intrinsically "safe." It simply means the extractives of the wood have little to no history in the scientific literature.

**KEY**

**Use Caution** = Some research indicates problematic extractives, but these are unlikely to move with water or at high-enough concentrations to be concerning.

**Avoid** = Likely problematic for sensitive populations

**Unsafe** = Assays or other tests have confirmed toxicity in some species.

**VERY Unsafe** = Toxicity is noted to cause problematic or even deadly responses.

| Name | Scientific Name | Safety | Research Notes | References |
|---|---|---|---|---|
| Alder, black | *Alnus glutinosa* | Use Caution | Contact dermatitis has been reported; confirmed with reported two patch tests. A tannin was suggested, but a different researcher found the tannin not to show results in a patch test. | Hausen 1981 |
| Birch* *Species specific, not widespread across genus | *Betula papyrifera* | Use Caution | Contact dermatitis has been reported in woodworkers, which was verified with patch tests. A quinone from fungi has been reported in the wood—not species specific. | Hausen 1981 |
| Camphor | *Cinnamomum camphor* | Very Unsafe | Headache, giddiness (from Hausen 1981). Noted toxicity in review by Chen. USDA restricts amounts in commercial products to no more than 11%. Known to cause seizures in children and death and can have similar (though less severe) effects in adults. | Hausen 1981, Chen et al. 2013<br>Khine et al. 2009<br>Love et al. 2004 |
| Canary wood | *Centrolobium tomentosum* | Unsafe | Considered mildly toxic; tested via ethanol extraction and brine shrimp bioassay. | Carlos de Sá Silva et al. 2013 |
| Cedar, incense | *Calocedrus decurrens* | Unsafe | Contact dermatitis reported; contact eczema confirmed with patch tests. Contains thymoquinol and thymoquinone (a strong sensitizer). | Hausen 1981 |
| Cedar, western red | *Thuja plicata* | Unsafe | Allergic contact dermatitis (even when finished); respiratory issues; contains thymoquinone and other extractives. Sensitization seen in guinea pig patch tests. | Bleumink et al. 1973;<br>Hausen 1981, 2012;<br>Huilaja et al. 2016 |
| Douglas fir | *Pseudotsuga menziesii* | Avoid | Allergic reaction reported in 2/82 people. Splinters reported to heal slowly; some patch tests reported resins containing terpenes a-pinene, thunbergol. | Hausen 1981 |
| Ebony, Ceylon | *Diosyros ebenum* | Unsafe | Eczema of skin reported; severe nose mucosa from sawdust; contact dermatitis reported for other ebony species; contain quinones; patch test reported. | Hausen 1981 |
| Hickory, shagbark | *Carya ovata* | Use Caution | Contains small amounts of juglone, the problematic extractive of black walnut. | Soderquist 1973 |
| Limba | *Terminalia superba* | Avoid | Wounds with splinters inflame and heal slowly. Reported as causing contact dermatitis; positive epicutaenous test reported; however, Hausen found no evidence of sensitizing capacity, quinones, or alkaloids that might cause issues. | Hausen 1981, 2012 |

| | | | | |
|---|---|---|---|---|
| Mahogany, African | *Khaya anthotheca* | Unsafe | Anthothecol has been found to be a sensitizer in the wood; others are likely. 2,6-dimethoxy-1,4-benzoquinone has been reported in *K. anthotheca* and *K. ivorensis* and is a possible sensitizer. | Hausen 1981 |
| Mahogany, Cuban | *Swietenia mahagoni* | Unsafe | Contact dermatitis reported. Contains a sensitizer similar to anthothecol; an enthanol extract of the wood induces sensitivity in guinea pigs. 2,6-dimethoxy-1,4-benzoquinone reported in the wood. | Hausen 1981 |
| Pao ferro (santos rosewood) | *Machaerium* spp. | Avoid | Contact dermatitis (severe) reported in medical literature in several cases. | Bonny et al. 2013<br>Ferreira et al. 2011 |
| Olive wood | *Olea europaea* | Avoid | Contact dermatitis of bracelets, etc. Sensitizing, ethanolic extractions yielded quinone extractions, and sensitization experiments with guinea pigs showed sensitizing effects. | Hausen 1981 |
| Pine, radiata | *Pinus radiata* | Use Caution | Allergic contact dermatitis reported; however, sensitization experiment using an ethanol extraction showed only weak sensitizing capacity. | Hausen 1981 |
| Pine, Scots | *Pinus sylvestris* | Avoid | Contact dermatitis widely reported; patch tests with extractives and sawdust are reported by author; may have pinosylvin, coniferyl benzoate, or other terpenoids. | Hausen 1981 |
| Purpleheart | *Peltogyne densiflora* | Avoid | Dust causes cough; tested wood has been found to contain an allergen; strong smell | Woods and Calnan 1976 |
| Redwood | *Sequoia sumpervirens* | Unsafe | Reactions including difficulty breathing and tachycardia reported by sawyers. | Woods and Calnan 1976<br>Chan-Yeung and Abbound 1976 |
| Rosewood, Brazilian | *Dalbergia nigra* | Unsafe | Contains contact allergens, dalbergiones. Guinea pigs show moderate allergic reactions. | Hausen 1981 |
| Rosewood, East Indian | *Dalbergia latifolia* | Unsafe | Most dalbergia species toxic, irritants, sensitizers. "Fiddlers neck" seen in violin players. Dalbergiones such as latinone and Dihydro-2',4-dimethoxydalbergione. | Hausen 1981 |
| Samba | *Triplochiton scleroxylon* | Avoid | Inhalation of dust caused asthma in a sensitized individual. | Ferrer et al. 2001 |
| Sassafras | *Sassafras albidum* | Very Unsafe | Extractive safrole causes cancer in rats. Banned by FDA in commercial products in 1960. Causes liver damage. | Segelman et al. 1976 |
| Sissoo | *Dalbergia sissoo* | Use Caution | Eczema reported with skin contact (one confirmed report). | Gómez-Muga et al. 2009 |

| Teak | *Tectonia grandis* | Unsafe | Allergic reactions to the wood known. Lapachol may be one of the sensitizers; deoxylapachol seems to be stronger but both have been seen to have no effect on guinea pigs. Other quinones may be active. | Hausen 1981 |
|---|---|---|---|---|
| Tigerwood/Goncalo alves | *Astronium fraxinifolium* | Avoid | Reported to cause severe skin irritations; speculated to contain catechols or phenolic vesicants (in Hausen), though not backed up with specific sensitivity studies. | Hausen 1981 |
| Tulip tree / Whitewood | *Lirodendron tulipifera* | Use Caution | Sensitivity to the wood was reported and tested with patch tests; only fresh wood produced dermatitis. | Hausen 1981 |
| Walnut, American | *Juglans nigra* | Use Caution | Contact dermatitis reported; juglone extracted from heartwood found to be a strong sensitizer in guinea pigs and horses. | Hausen 1981<br>True and Lowe 1980 |
| Zebrawood | *Microberlinia brazzavillensis* | Use Caution | Dust reported to cause asthma, both immediate and late onset. | Bush et al. 1977 |

**Appendix I References**

Bleumink, E., J. C. Mitchell, and J. P. Nater. "Allergic Contact Dermatitis from Cedar Wood (*Thuja plicata*)." *British Journal of Dermatology* 88, no. 5 (1973): 499-504.

Bonny, M., O. Aerts, J. Lambert, and H. Lapeere. "Occupational Contact Allergy Caused by Pao Ferro (Santos Rosewood): A Report of Two Cases." *Contact Dermatitis* 68, no. 2 (2013): 126-28.

Bush, R. K., J. W. Yunginger, and C. E. Reed. "Asthma Due to African Zebrawood (Microberlinia) Dust." *American Review of Respiratory Disease* 117, no. 3 (1978): 601-603.

Carlos de Sá Silva, Roberto, Julio Cesar Raposo de Almeida, Ana Aparecida da Silva Almeida, Gokithi Akisue, Matheus Diniz Gonçalves Coelho, and José Roberto Pereira. "Evaluation of the Toxicity of Araribá (*Centrolobium tomentosum*) Using Brine Shrimp Test." *Ambiente e Agua: An Interdisciplinary Journal of Applied Sciences* 8, no. 4 (2013): 158-67.

Chan-Yeung, M., and R. Abbound. "Occupational Asthma Due to California Redwood (*Sequoia sempervirens*) Dusts." *American Review of Respiratory Disease* 114, no. 5 (1976): 1027-31.

Chen, Weiyang, Ilze Vermaak, and Alvaro Viljoen. "Camphor—a Fumigant during the Black Death and a Coveted Fragrant Wood in Ancient Egypt and Babylon—a Review." *Molecules* 18, no. 5 (2013): 5434-5454.

Ferreira, O., M. João Cruz, A. Mota, A. Paula Cunha, and F. Azezedo. "Erythema Multiforme-Like Lesions Revealing Allergic Contact Dermatitis to Exotic Woods." *Cutaneous and Ocular Toxicology* 31, no. 1 (2011): 61-63.

Ferrer, A., F. Marañón, M. Casanovas, and E. Fernández-Caldas. "Asthma from Inhalation of *Triplochiton scleroxylon* (Samba) Wood Dust." *Journal of Investigational Allergology & Clinical Immunology* 11, no. 3 (2001): 199-203.

Gómez-Muga, S., J. A. Ratón-Nieto, and I. Ocerin. "An Unusual Case of Contact Dermatitis Caused by Wood Bracelets." *Contact Dermatitis* 61, no. 6 (2009): 351-52.

Hausen, Björn M. *Woods Injurious to Human Health: A Manual*. Berlin and New York: Walter de Gruyter, 1981.

Hausen, B. M., G. Bruhn, and W. A. Koenig. "New Hydroxyisoflavans as Contact Sensitizers in Cocus Wood *Byra ebenus* DC (Fabaceae)." *Contact Dermatitis* 25, no. 3 (1991): 149-55.

Huilaja, L., M. E. Kubin and R. Riekki. "Contact Allergy to Finished Woods in Furniture and Finishings: A Small Allergic Contact Dermatitis Epidemic to Western Red Cedar in Sauna Interior Decoration." *Journal of the European Academy of Dermatology and Venereology* 30, no. 1 (2016): 57-59.

Soderquist, Charles J. "Juglone and Allelopathy." *Journal of Chemical Education* 50, no. 11 (1973): 782-83.

Woods, Brian, and C. D. Calnan. "Toxic Woods." *British Journal of Dermatology* 94, no. 13 (1976): 1-97.

"Woods [MAK Value Documentation, 1999]." In *The MAK-Collection for Occupational Health and Safety: Annual Thresholds and Classifications for the Workplace*. Edited by Helmut Greim, 284-89. Weinheim, Germany: Wiley-VCH Verlag, 2012.

"Woods: *Terminalia ivorensis* A. Chev. [MAK Value Documentation 2002]." In *The MAK-Collection for Occupational Health and Safety*. Edited by Helmut Greim, 259-62. Weinheim, Germany: Wiley-VCH Verlag, 2012.

# Appendix II

## LIST OF SELLERS OF WOODEN TOYS

The following is a list of sellers of wooden toys, both commercial and small, that are known to currently (or have in the very recent past) used *teething-safe woods* or are willing to substitute teething-safe woods for nonteething safe woods in their products. Please remember that most woods are completely fine for hand play, and as such, this list is specific for those interested in products for young children. The list also generally focuses on US toymakers, or production lines that can be purchased in the US.

**Barclay Wood Toys and Blocks.** www.barclaywoods.com

**Brio.** www.brio.us

**Colorado Goods and Wood.** www.etsy.com/shop/coloradogoodsandwood#items

**Community Playthings.** www.communityplaythings.com

**Fagus.** http://fagus-holzspielwaren.de/

**From Jennifer.** www.etsy.com/market/from_jennifer

**Grimm.** www.grimms.eu/en/

**Haba.** www.habausa.com

**Kapla.** www.kaplaus.com

**Mirus Toys.** www.etsy.com/shop/MirusToys

**Nova Naturals.** www.novanatural.com

**Paci Catchers.** www.pacicatchers.com

**SensoryPlay.** www.etsy.com/shop/sensoryplay

**T.C.'s Wooden Toys.** www.tcswoodentoys.com

**Tree Hopper Toys.** www.treehoppertoys.com

**Two Raccoon Hollow.** www.facebook.com/TwoRaccoonHollow

**Uncle Goose.** https://unclegoose.com

# Appendix III

## QUICK AND DIRTY WOOD ID

While this book is not meant as a primer on wood identification (for such a book, see appendix IV, "Next Steps"), following are some macro and corresponding micro images of some common woodworking woods. The macro sections should be visible with a 10x hand lens and a bit of sanding to the end grain. The micro is a microscopic view, usually 100x.

Included are labels of the relevant ID characteristics, such as vessel/tracheid distribution, parenchyma distribution, resin canals, spiral thickenings, etc. It's not important so much that you know what these elements are in quick and dirty ID, but more that you learn to recognize them and pair them together so you can tell different wood species apart—if only enough to determine which are appropriate for the job you're trying to do.

Because many softwoods must be identified on their *radial* plane instead of their *transverse* plane, some images show both a 100x and 400x image of the radial plane so that the *cross-field pitting* shape can be observed. It is this area, where the ray cells bisect the longitudinal tracheids and simple pits meet bordered pits, in which unique pit shapes occur and aid in ID (for more information on wood ID, cross-field pits, etc., please see appendix IV, "Next Steps").

Along with each image are included the common name(s), scientific name, regions found in, relevant macro planes and micro planes, and relevant ID information.

# HARDWOODS

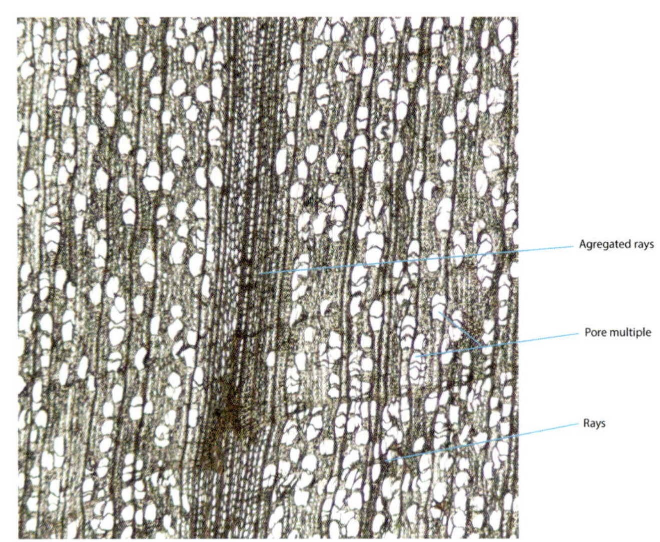


alder, European (*Alnus glutinosa*): Throughout Europe, parts of Asia and Africa. Alder transverse plane shows multiple pores and rays. Pay special attention to the rays, since they can be present as single ray strands, aggregated rays, or both.

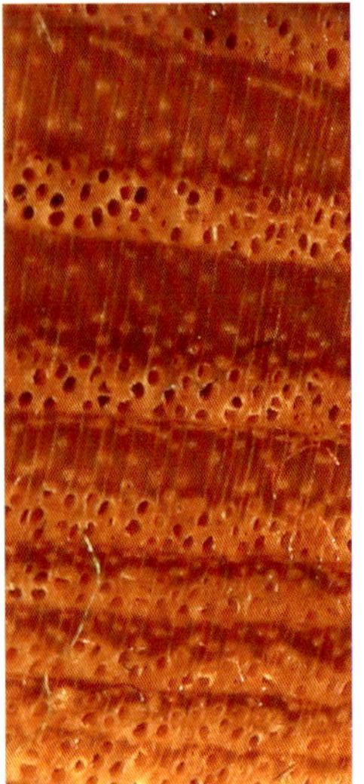

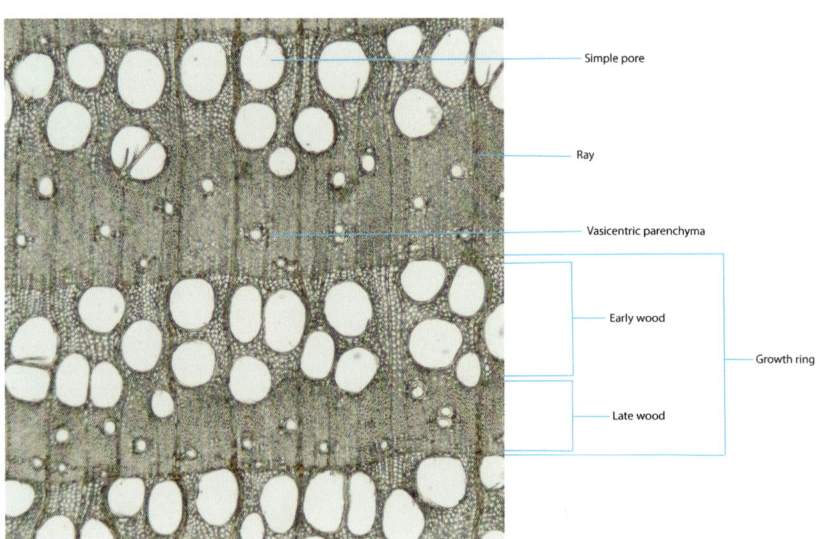


ash, black (*Fraxinus nigra*): Eastern Canada and eastern US. Black ash transverse plane. The wood is ring porous, with two distinct rows of large pores. The transverse plane also shows vasicentric parenchyma around the latewood pores.

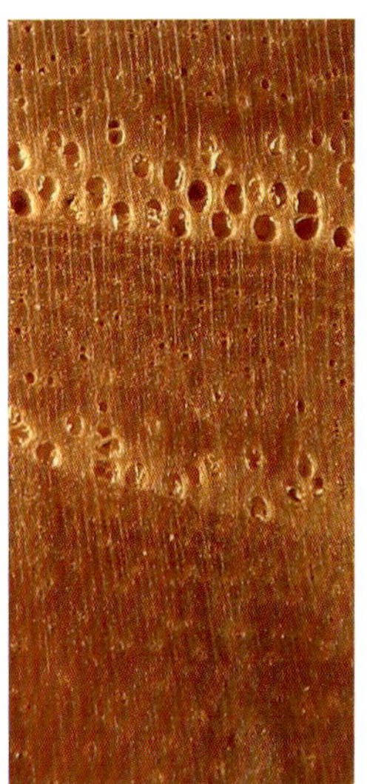

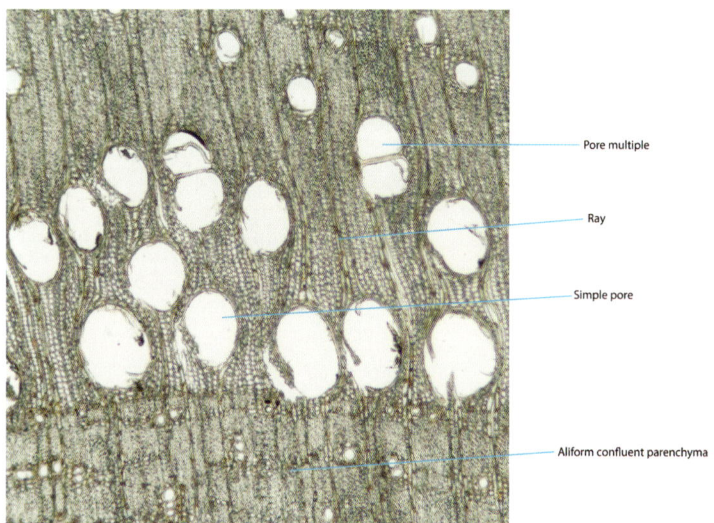


ash, white (*Fraxinus americana*): Eastern and central Canada and US. White ash transverse plane shows simple and multiple pores. The wood is ring porous and can have either two or three rows of earlywood. Latewood shows pores connected by aliform-confluent parenchyma.

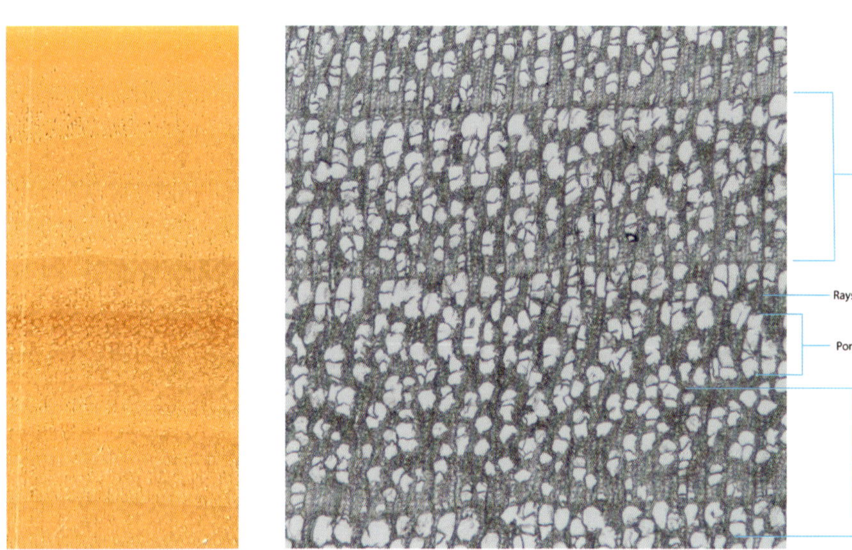


aspen (*Populus* sp.): Widely distributed throughout the Northern Hemisphere. Transverse plane highlights the diffuse pore multiples.

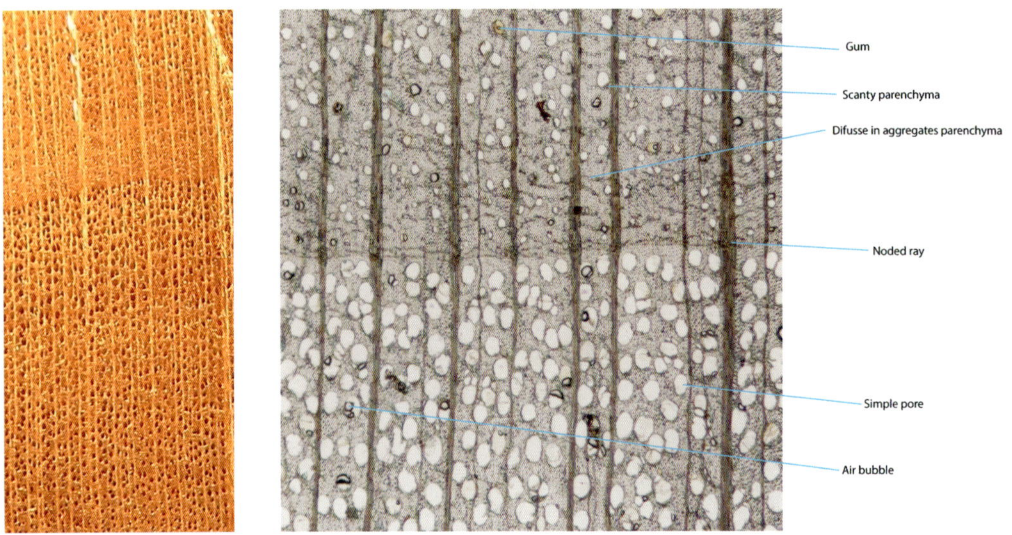


beech, American (*Fagus grandifolia*): Eastern US and southeastern Canada. Transverse plane shows characteristic nodded rays (rays balloon out when crossing the growth ring boundary). This species is semi-ring porous and has predominant simple pores, some of which contain yellow gum.

beech, European (*Fagus sylvatica*): Throughout Europe and part of the Mediterranean. Transverse plane showing nodded rays (thicker at the growth ring), diffuse in aggregate parenchyma, and predominantly simple pores.

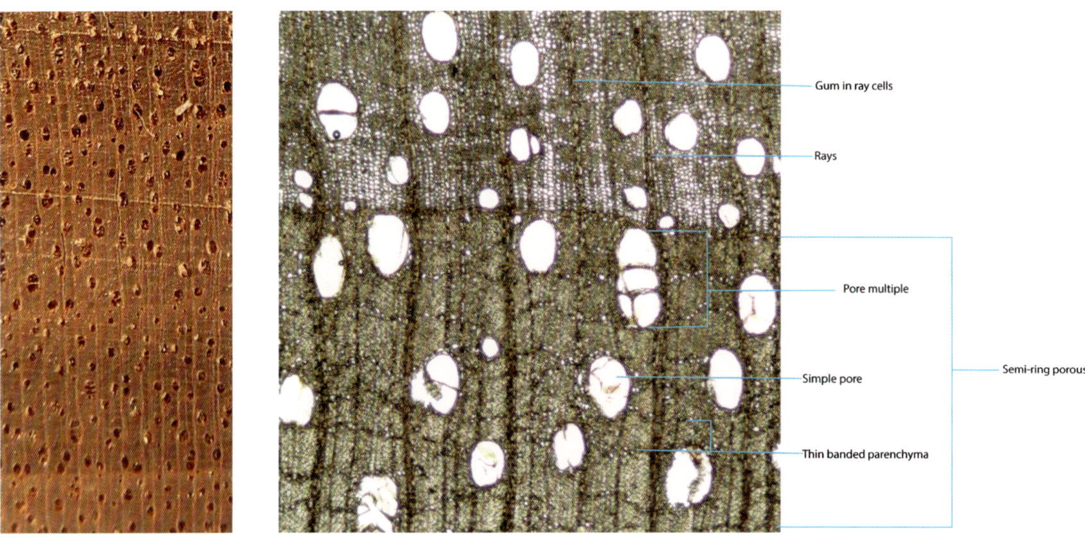


black walnut (*Juglans nigra*): Eastern North America. Transverse plane shows simple and multiple pores, thin-banded parenchyma, and semi-ring porous growth ring.

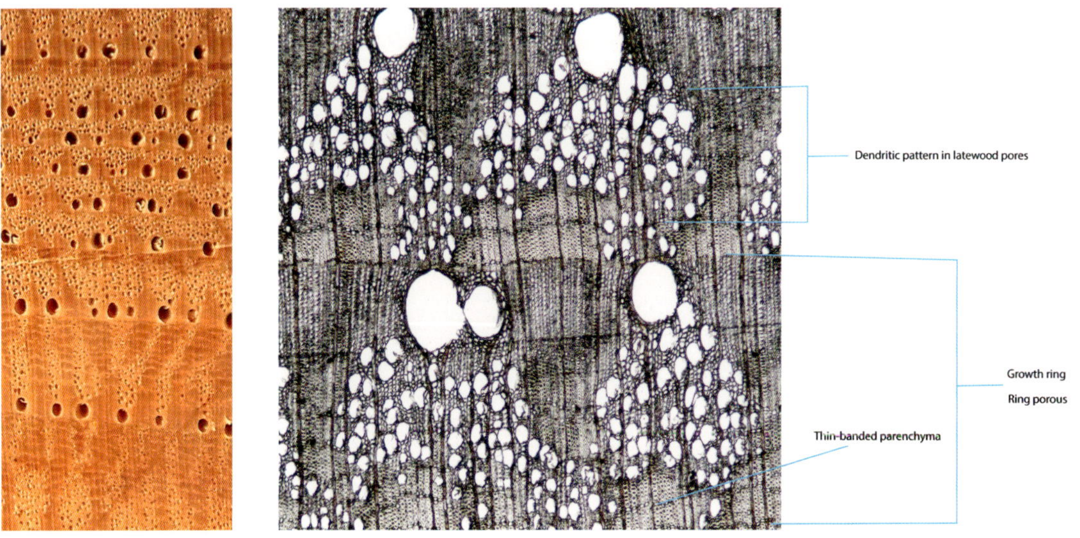


chinkapin (*Chrysolepis chrysophylla*): Western US. Transverse plane shows ring porosity. Latewood shows a dendritic (flame-like) pore pattern, which is characteristic for this species. Thin-banded parenchyma (apotracheal) are also easy to observe.

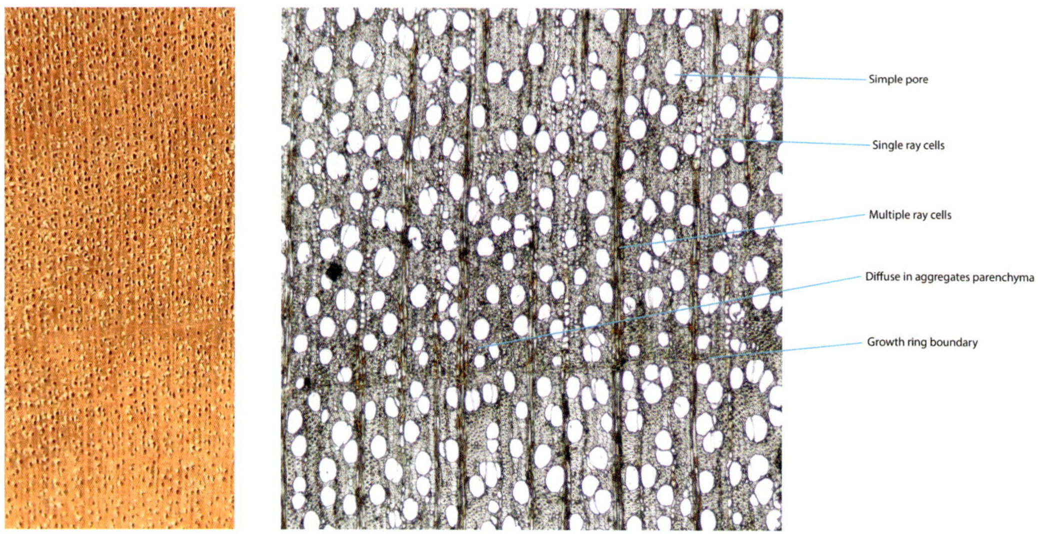


dogwood (*Cornus florida*): Eastern North America. The transverse plane predominantly shows simple pores, two distinct ray arrangements (single and multiple), diffuse in aggregates parenchyma (apotracheal parenchyma), and a growth ring boundary of two to three cells of width.

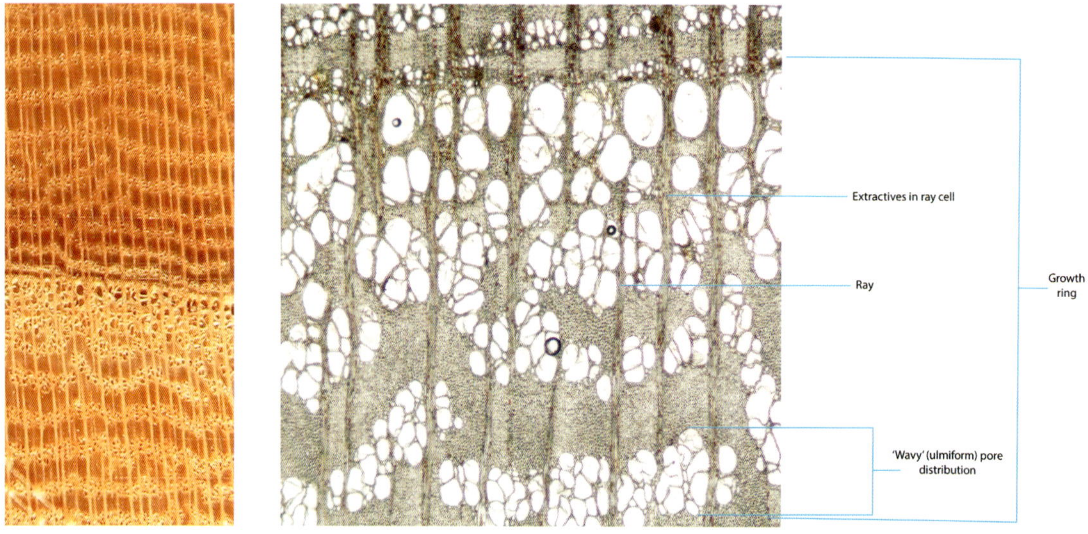


elm, American (*Ulmus americana*): Eastern North America. Transverse plane shows an ulmiform (wavy) pore distribution (identification characteristic in elm species), with pore bands raging from two to four pores. The ray cells show red-brown extractives.

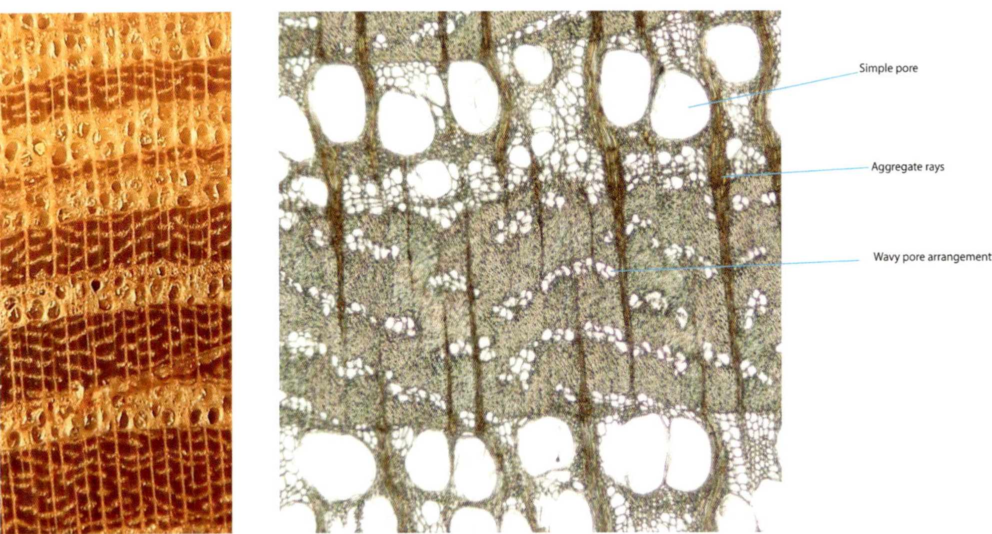


elm, red (*Ulmus rubra*): Eastern North America. Transverse plane shows the ulmiform (wavy) pore arrangement. Red elms have more solitary pores than American elm, and more than one row of earlywood.

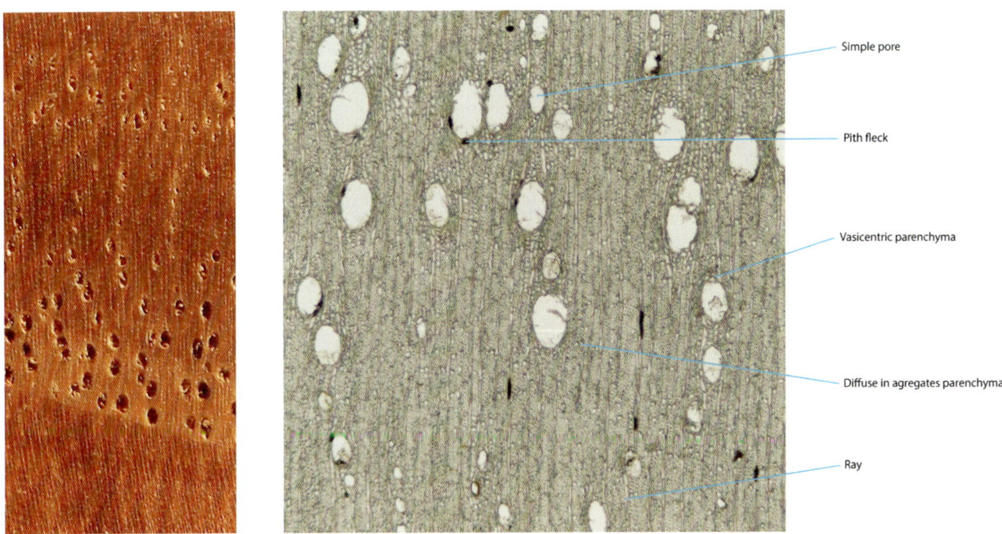


hickory (*Carya* sp.): Various species range throughout the world. The transverse plane shows simple pores. In the image it is possible to differentiate two types of parenchyma: vasicentric (paratracheal) and diffuse in aggregates (apotracheal). Some of the ray cells and parenchyma show the presence of pith fleck (a dark streak common to many North American hardwoods, caused by an insect).

madrone (*Arbutus menziesii*): Western North America. The transverse plane shows diffuse simple pores. The growth ring boundary is evident. Madrone has *no* longitudinal parenchyma cells.

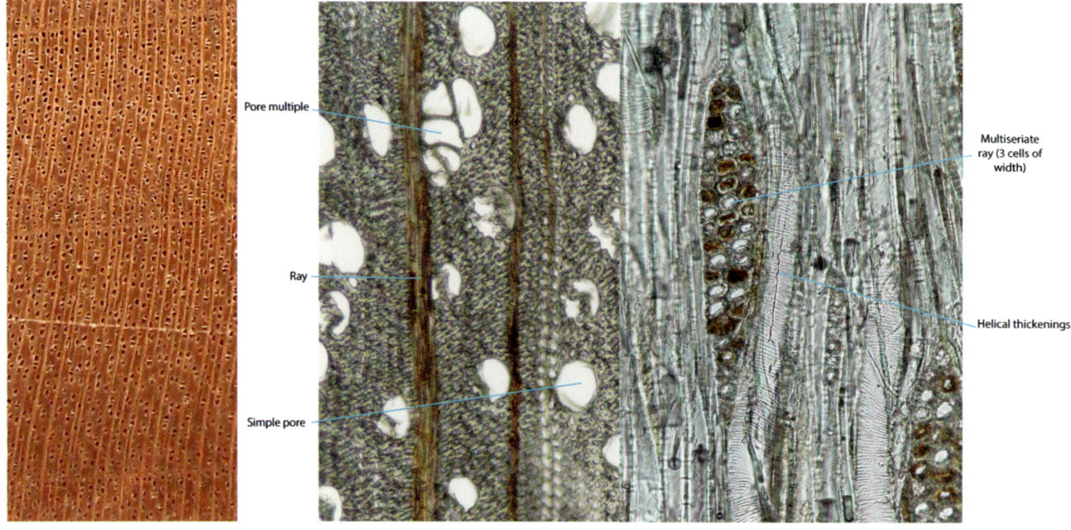


maple, big leaf (*Acer macrophyllum*): Western North America. Transverse plane (*left callout*) shows simple and multiple pores. The tangential section (*right callout*) shows ray seriation of three cells (3 rays cells of width), and helical thickenings that are characteristic for the genus *Acer*. Like most soft maples, it has smaller rays (fewer seriate), and generally the rays are of roughly the same size. This is in contrast to hard maples (such as sugar maple), which have rays of two distinct sizes, usually uniseriate and then a multiseriate.

maple, sugar (*Acer saccharum*): Eastern and central Canada and US. Transverse plane (*left callout*) shows simple and multiple pores. The tangential section (*right callout*) shows ray seriation of five to seven cells (5–7 rays cells of width), and helical thickenings that are characteristic for the genus *Acer*. Hard maples (such as sugar maple) are characterized by having two distinct ray sizes, the larger five to seven seriate, and also uniseriate, which is key to distinguishing them from soft maples.

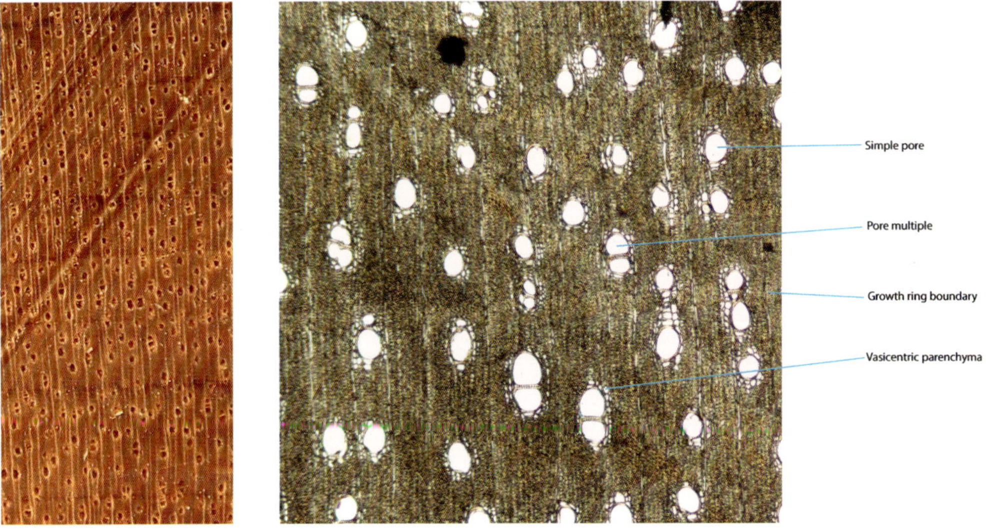


myrtle (*Umbellularia californica*): Coast side of California and Oregon, US. The transverse plane shows simple and multiple pores (predominantly in pairs). Special focus should be given to the vasicentric parenchyma cells (paratracheal parenchyma around the pores).

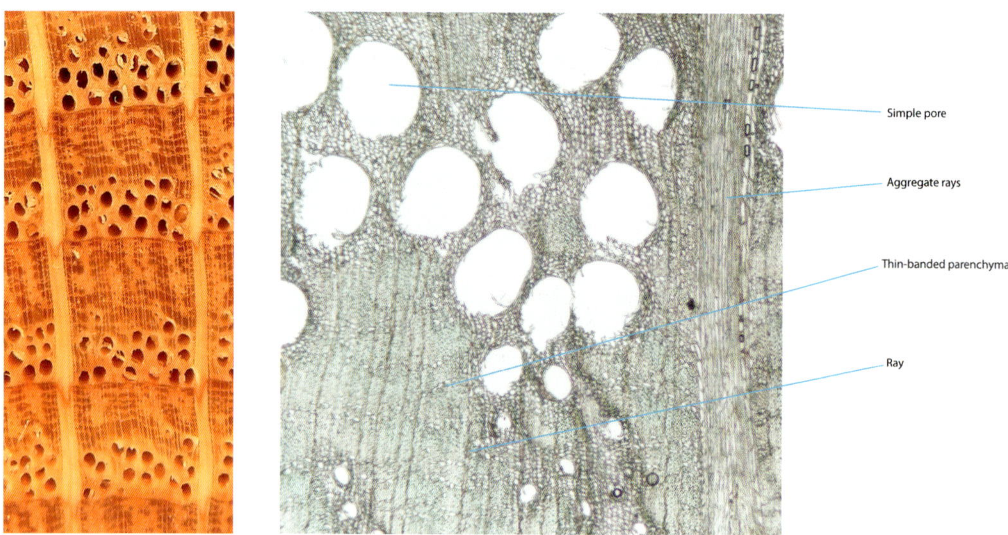


oak, red (*Quercus rubra*): Eastern and central US and Canada, with some introduction into Europe. The transverse plane shows simple pores, thin-banded parenchyma, and aggregate rays. A ring porous genus, red oak can be distinguished from white oak due to its relatively few to no tyloses, and more scattered latewood.

oak, white (*Quercus alba*): Eastern and central US and Canada. The transverse plane shows simple pores, tyloses (the shiny stuff in the pores), and ring-porous growth rings. White oaks can be distinguished from red oaks due to their abundant tyloses and dendritic latewood (forms a sort of smoke pattern in the latewood).

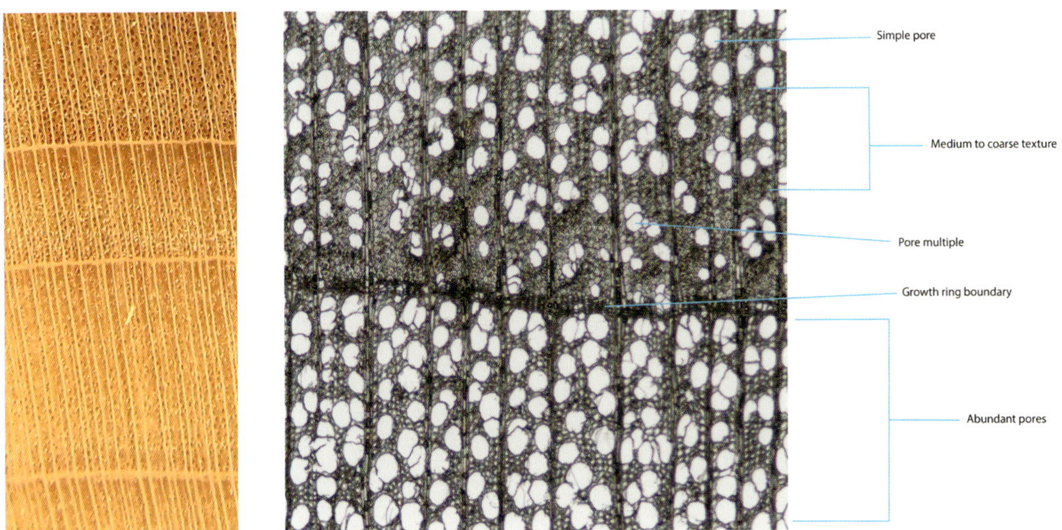


poplar (*Populus* sp.): Common throughout the Northern Hemisphere. The transverse plant shows a medium to coarse texture, abundant simple and multiple pores, and a distinct growth ring.

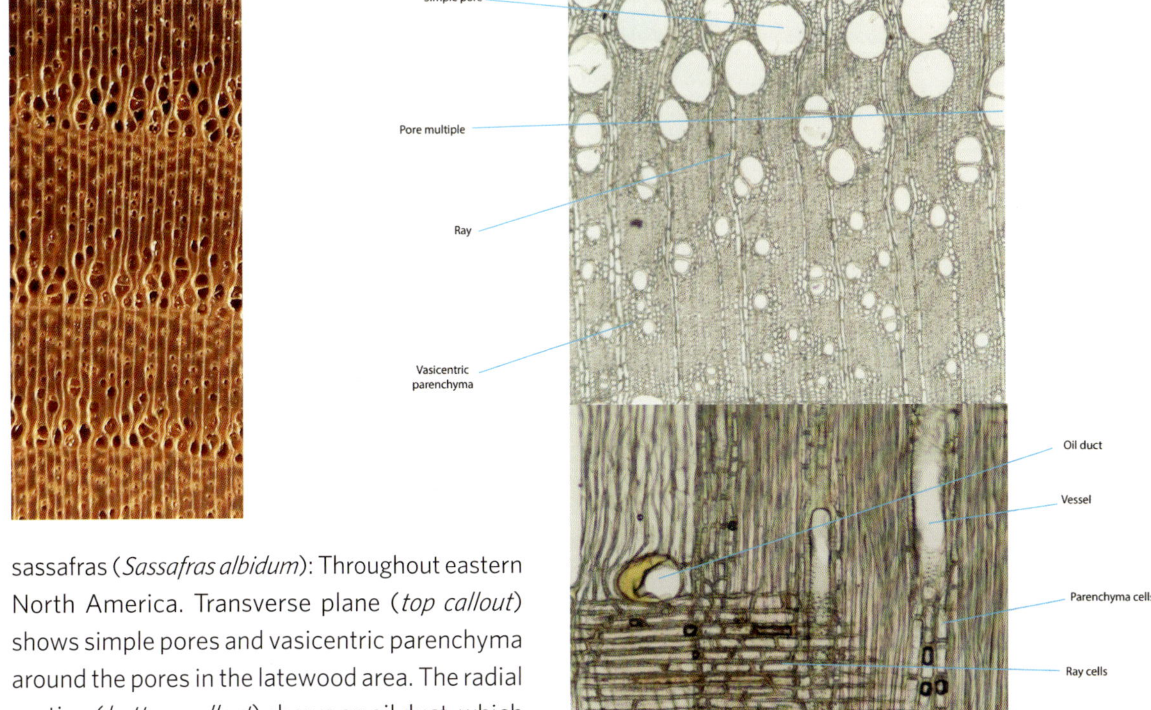


sassafras (*Sassafras albidum*): Throughout eastern North America. Transverse plane (*top callout*) shows simple pores and vasicentric parenchyma around the pores in the latewood area. The radial section (*bottom callout*) shows an oil duct, which is characteristic for this species.

# SOFTWOODS

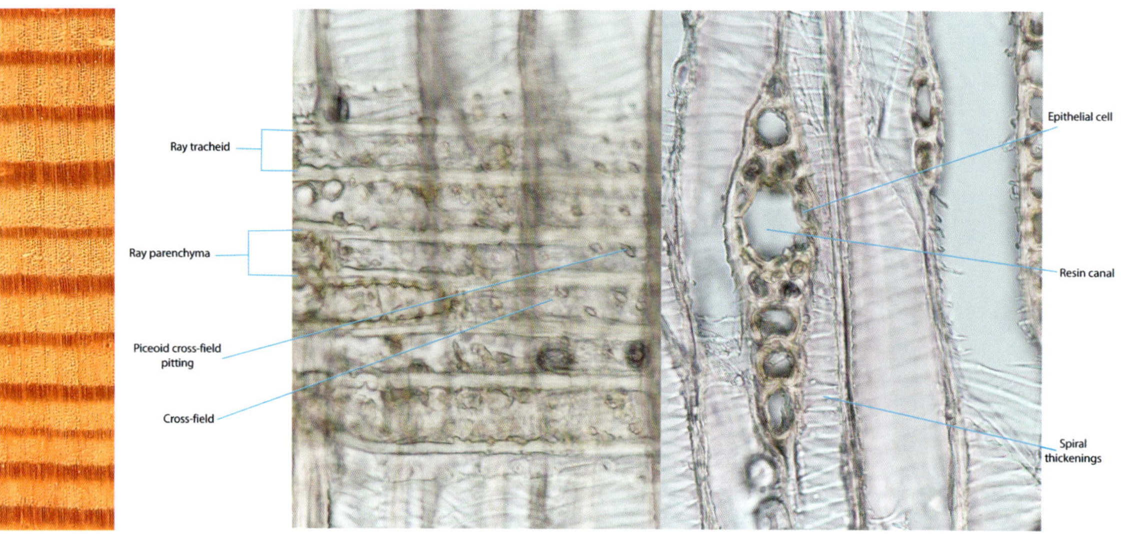


Douglas fir (*Pseudotsuga menziesii*): Western North America. Radial section (*left callout*) shows piceoid cross-field pitting. The tangential section (*right callout*) shows the fusiform ray, containing epithelial cells and a resin canal. In this section it is also possible to observe the spiral thickenings in the tracheids. All these identifying characteristics are unique for Douglas fir.

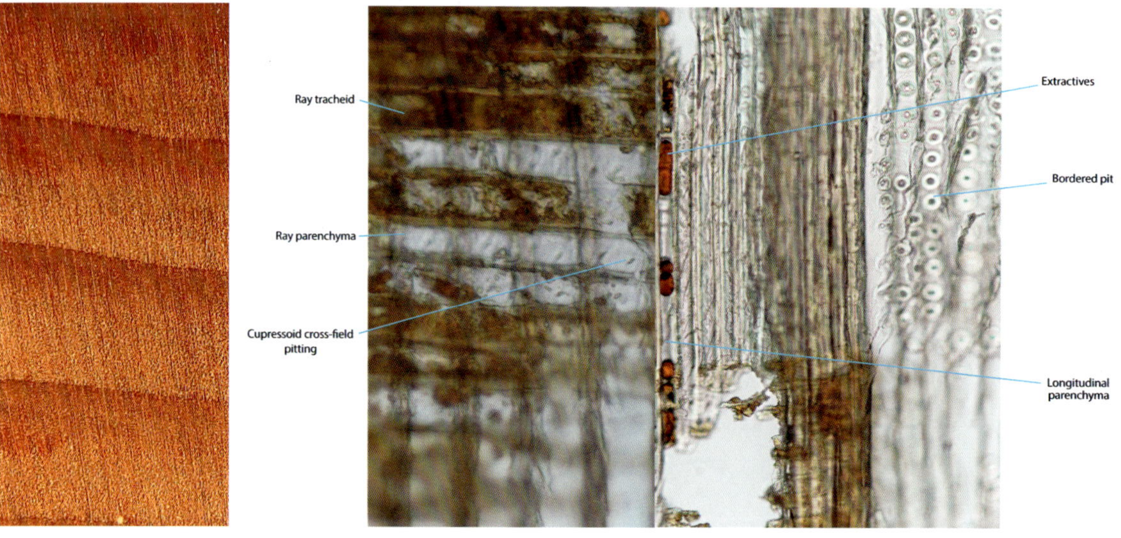


eastern red cedar (*Juniperus virginiana*): Eastern and edging into central North America. Radial section (*left callout*) shows cupressoid cross-field pitting. This species also show longitudinal parenchyma containing extractives.

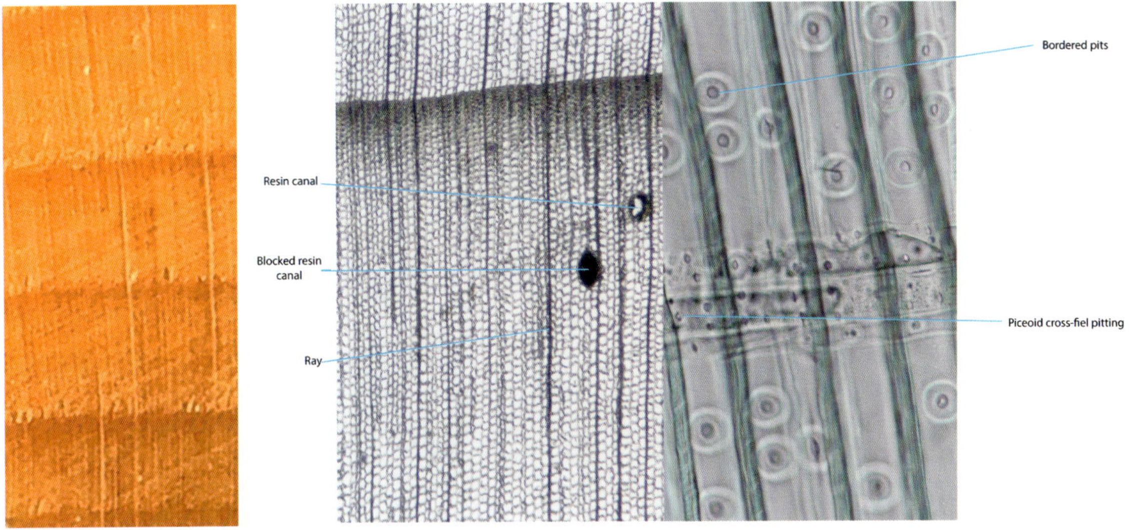


eastern spruce (*Picea rubens*): Eastern North America. Transverse plane (*left callout*) shows resin canals (one is blocked). The transverse plane (*right callout*) shows piceoid cross-field pitting.

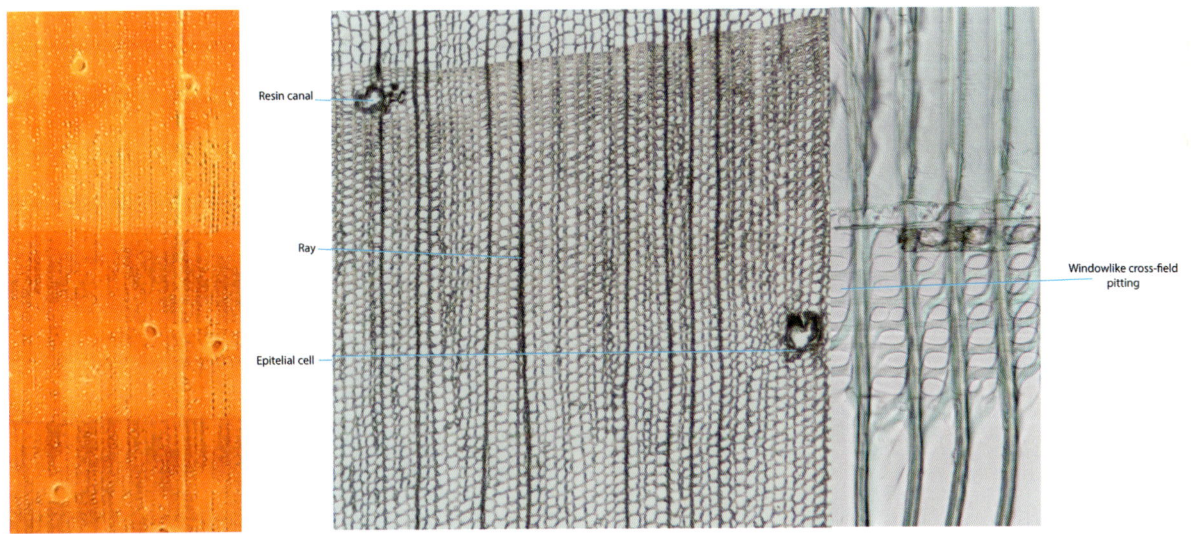


eastern white pine (*Pinus strobus*): Eastern North America. Transverse plane (*left callout*) shows resin canals. The radial section (*right callout*) shows window-like cross-field pitting.

lodge pole pine (*Pinus contorta*): Western North America. The transverse plane of lodge pole pine shows resin canals and a gradual transition from latewood to earlywood. Radial section shows dentate ray tracheids (characteristic of hard pines), and pinoid cross-field pitting.

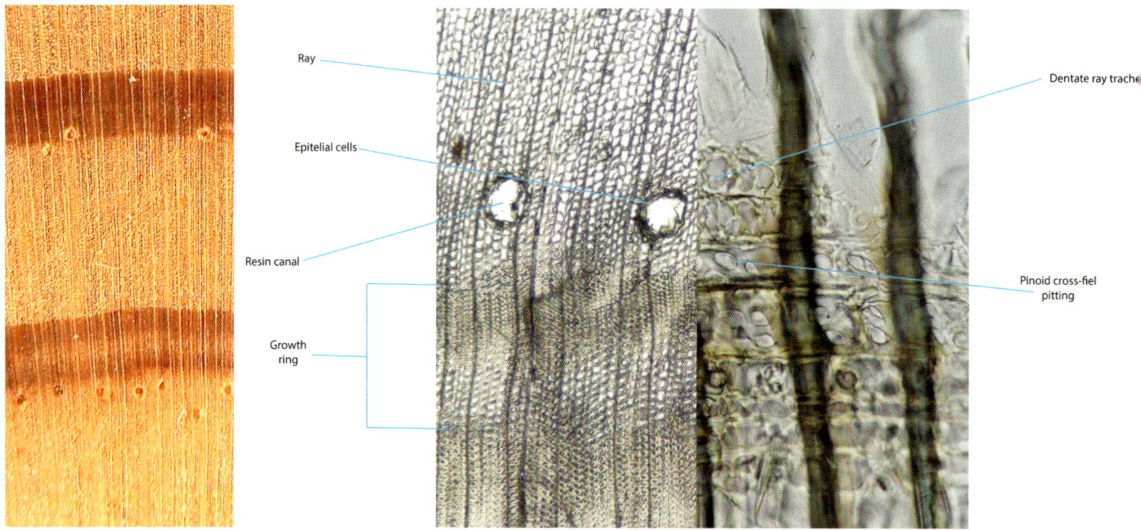


long leaf pine (*Pinus palustris*): Southeastern United States. Transverse plane (*left callout*) shows several resin canals formed by epithelial cells. The radial section (*right callout*) shows dentate ray tracheids (characteristic for hard pines), and pinoid cross-field pits.

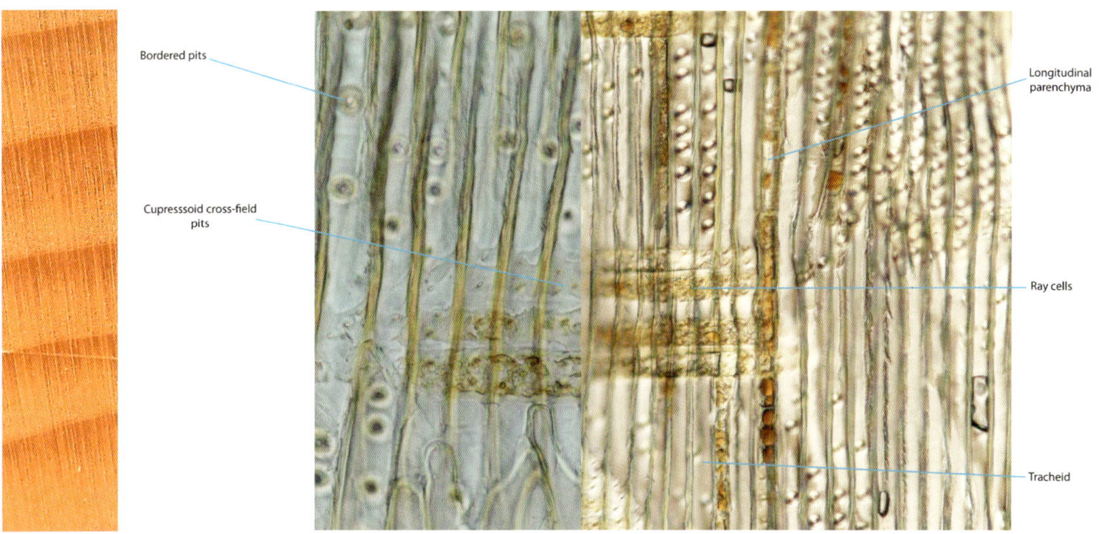


mountain hemlock (*Tsuga mertensiana*): Western North America. The radial plane shows the distinct longitudinal parenchyma. The zoomed-in radial plane shows the cupressoid cross-field pits indicative of the species.

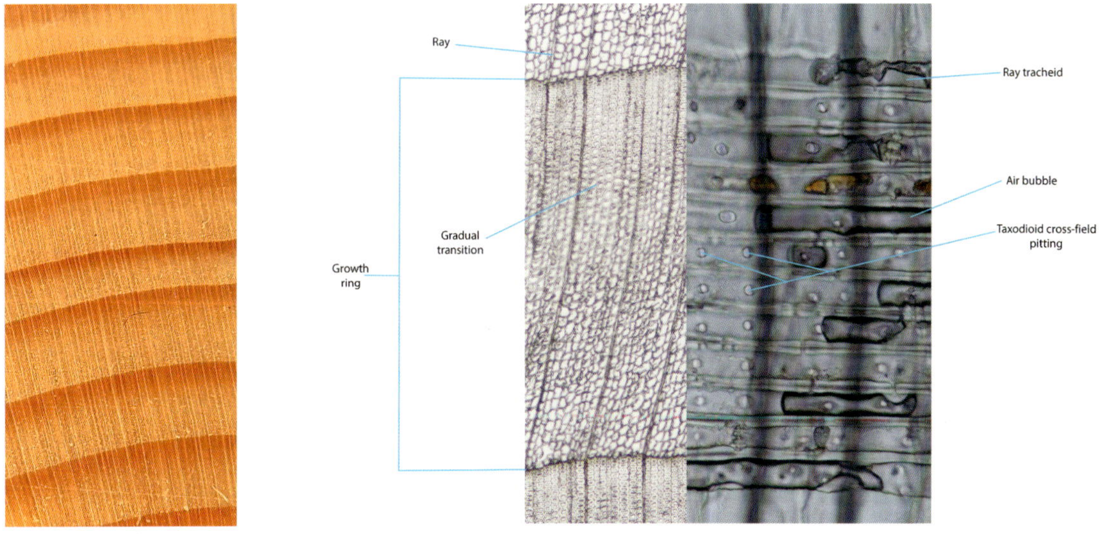


noble fir (*Abies procera*): Very limited range in California, Oregon, and Washington, primarily near the Cascade and Coast Range. Cross section (*left callout*) shows gradual transition from earlywood to latewood. The radial section (*right callout*) shows taxodioid cross-field pitting.

Pacific silver fir (*Abies amabilis*): Pacific Northwest of North America. Radial section shows taxodioid cross-field pitting. Ray tracheids are absent, which is an identification characteristic for this species.

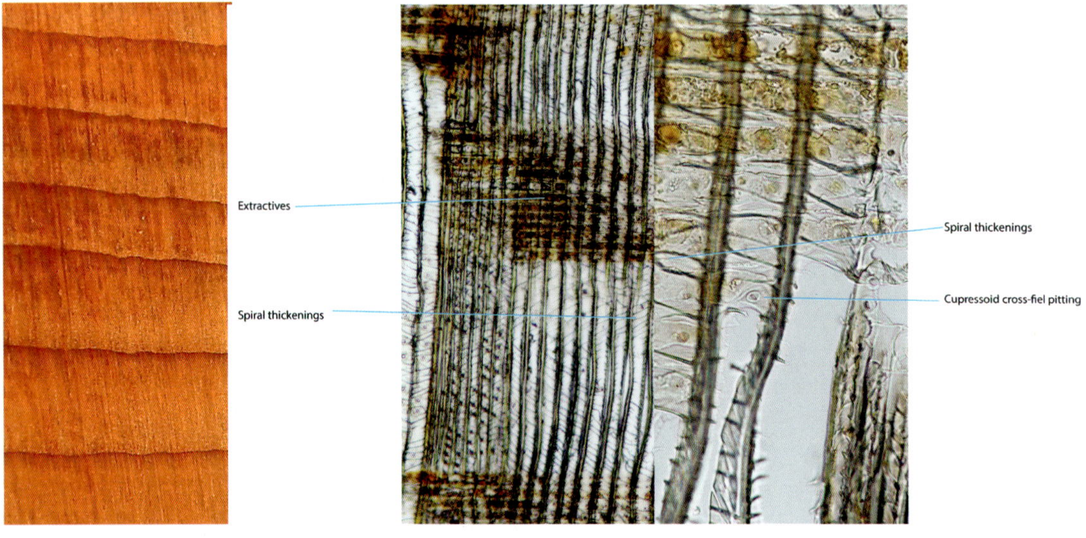


Pacific yew (*Taxus brevifolia*): Pacific Northwest of North America. Radial section shows presence of extractives in rays, spiral thickenings, and cupressoid cross-field pitting. Pay special attention to the lack of "normal" resin canals, which distinguish Pacific yew from Douglas fir (although any softwood can have resin canals due to trauma).

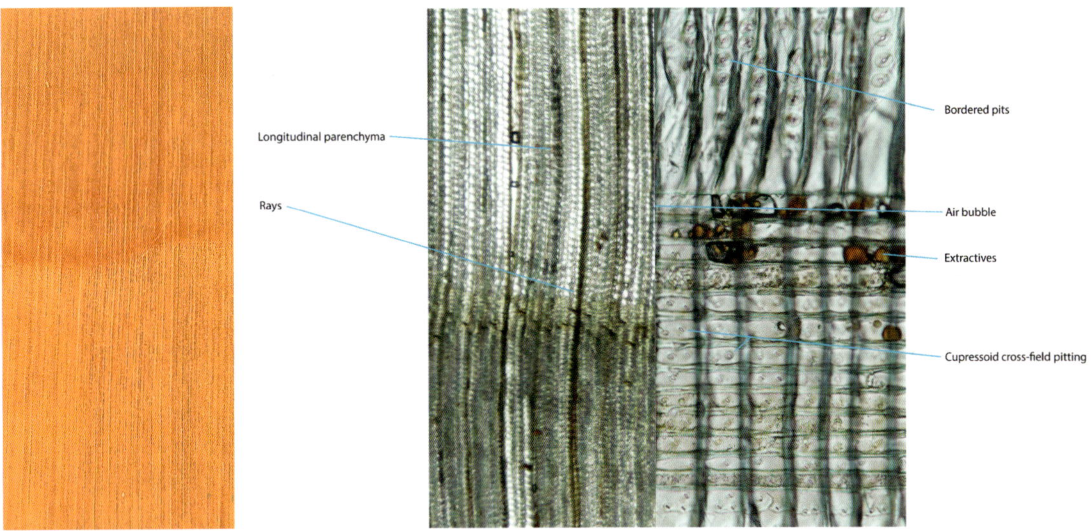


Port Orford cedar (*Chamaecyparis lawsoniana*): Very limited distribution across Oregon and California, US. The transverse plane shows longitudinal parenchyma cells. The radial section shows cupressoid cross-field pitting. It is important to notice the absence of ray tracheids, since these are characteristic for the species.

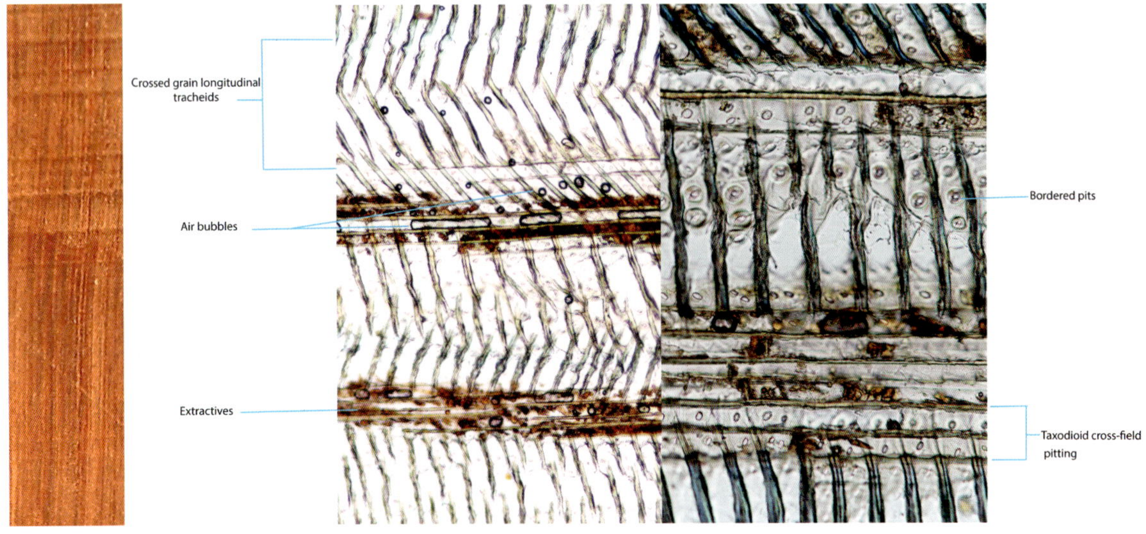


redwood (*Sequoia sempervirens*): Limited range across coastal California and Oregon, US. *On the left callout*, the radial section shows a detail of cross-grain and extractives in the rays, while the *right callout* shows the taxodioid cross-field pitting.

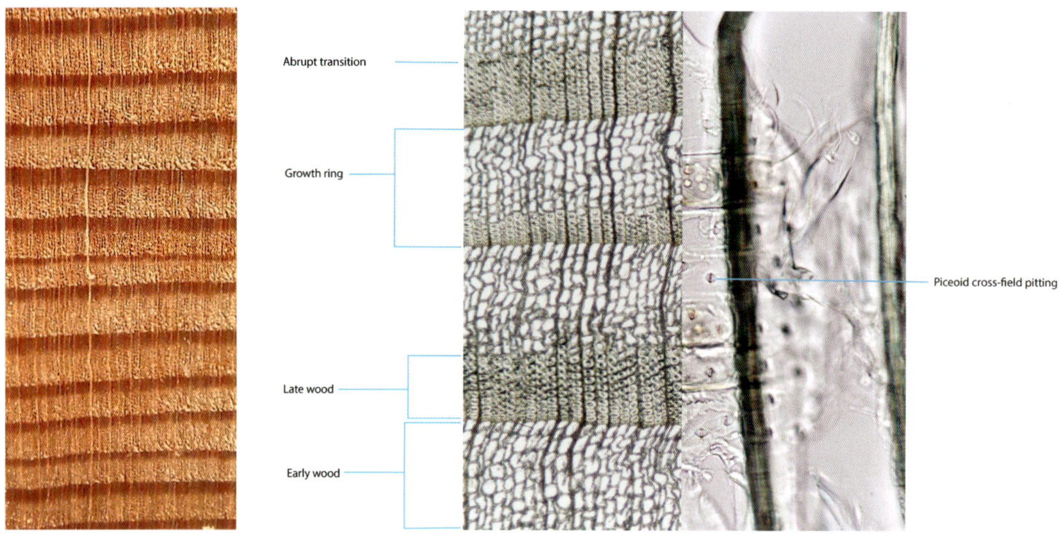


western larch (*Larix occidentalis*): Western and edging into central North America. The transverse plane shows an abrupt transition. The radial plane shows piceoid pitting in the ray crossing area.

## TROPICAL WOODS

(a mix of hardwoods and softwoods that grow in tropical climates and therefore have some unique ID characteristics)

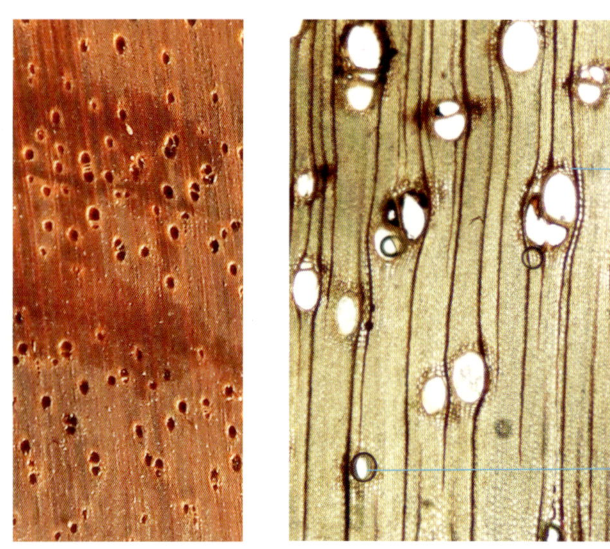

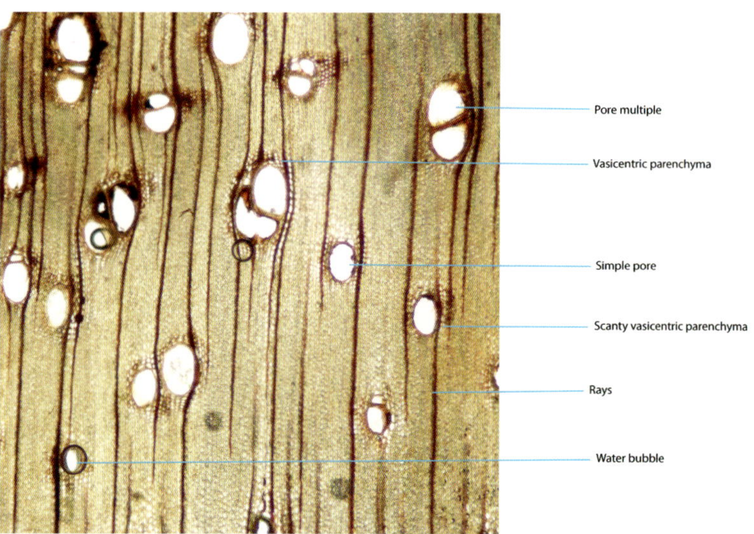


acacia (*Acacia* sp.): Throughout Australia and parts of Southeast Asia. The transverse plane shows simple and multiple pores. This species has two types of paratracheal parenchyma: vasicentric and scanty vasicentric (partial parenchyma around the pore).

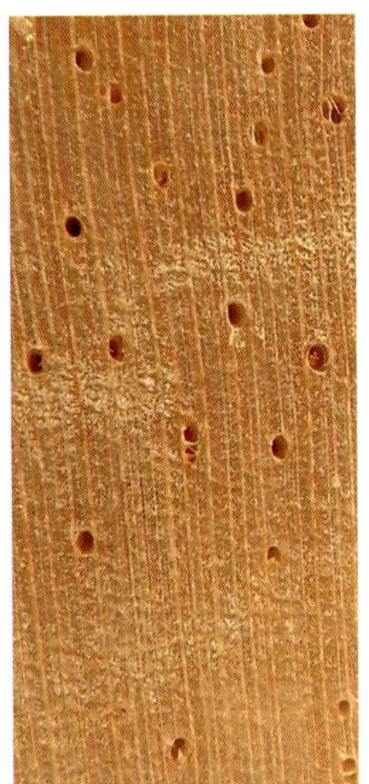

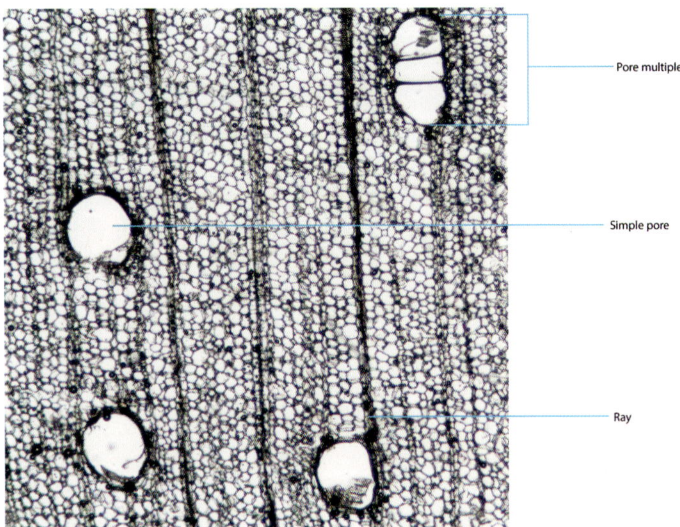


balsa (*Ochroma pyramidale*): Throughout sections of South America and Central America and into parts of Mexico. Transverse plane shows simple and multiple pores. The wood is diffuse porous (like most tropical hardwoods), though has wide distances between the vessel elements, which is a key ID trait.

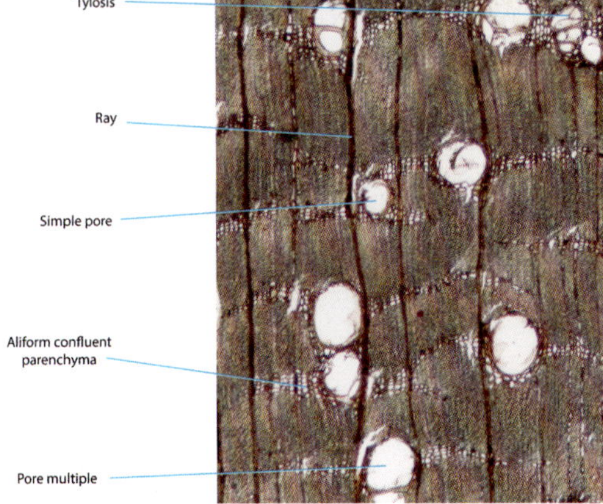


bloodwood (*Brosimum rubescens*): Throughout South America. Transverse plane (*left callout*) shows simple and multiple pores, tyloses, and aliform confluent parenchyma. The tangential section (*right callout*) shows gum ducts within in the ray cells.

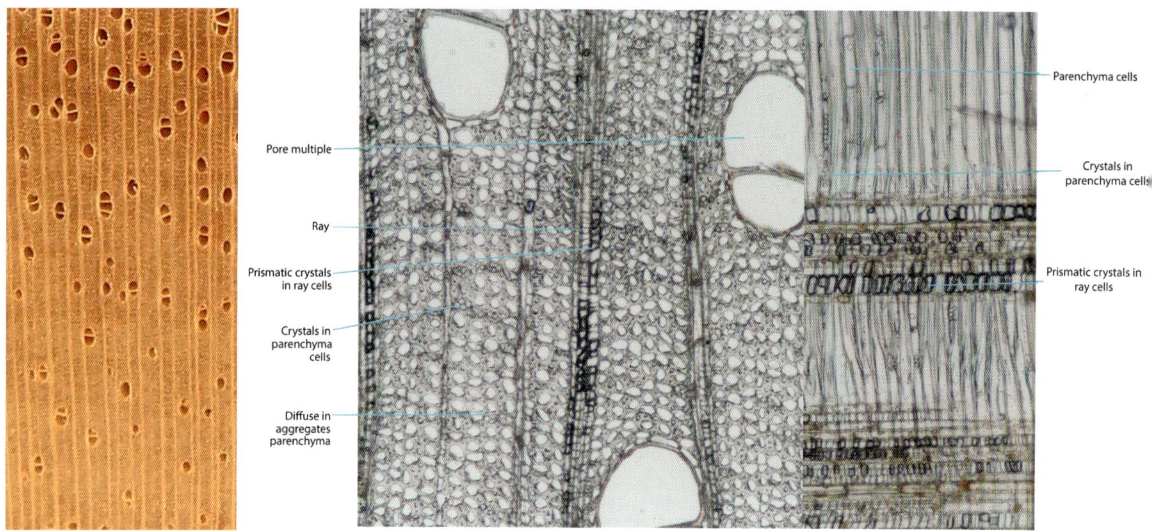


bolaina (*Guazuma crinita*): Eastern South America. Transverse plane (*left callout*) shows pores in pairs of two. The ray cells show visible prismatic crystals. Small chains of diffuse-in-aggregate parenchyma are also easy to identify, especially due to the presence of small, round crystals in them. The radial section (*right callout*) shows the longitudinal parenchyma with several rounded crystals inside, as well as prismatic crystals occupying ray cells.

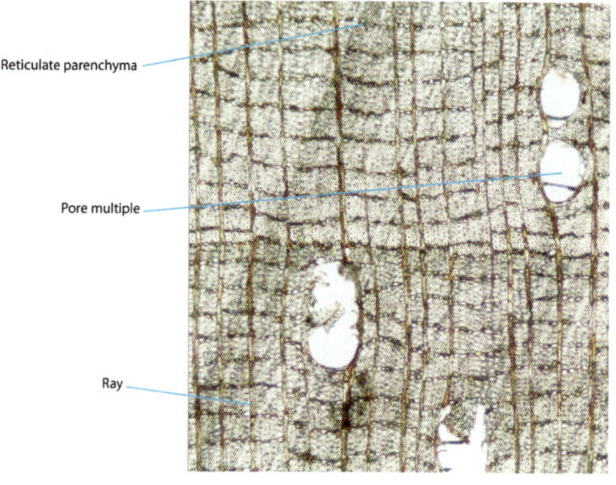


cachimbo (*Cariniana domestica*): Northern and central Brazil, parts of Peru. The transverse plane (*top callout*) shows pores in pairs. The reticulate parenchyma are one of the main characteristics in this species. The radial section (*bottom callout*) shows crystal inclusions in the parenchyma cells.

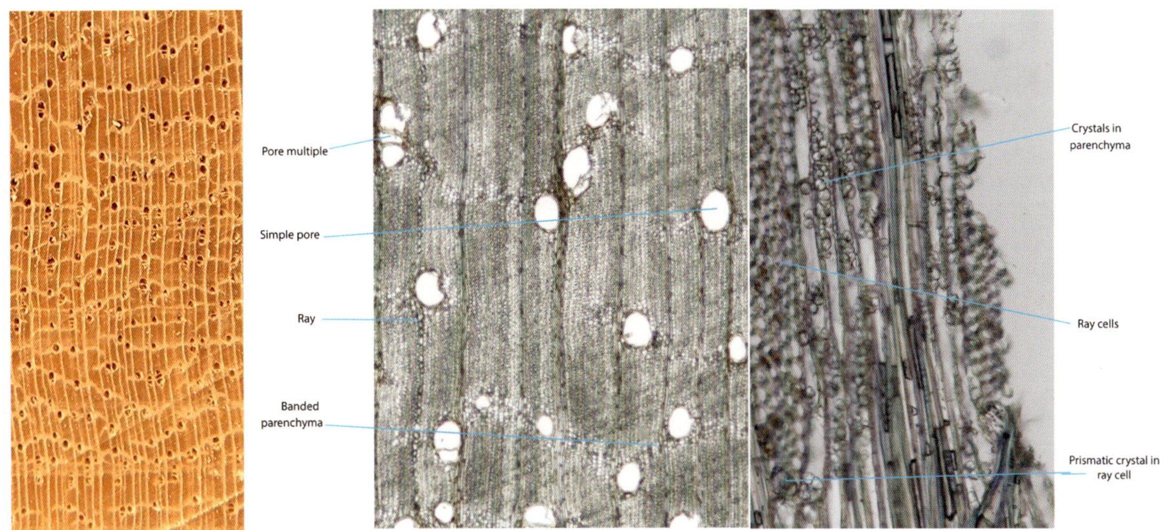


congona (a.k.a. breadnut) (*Brosimum alicastrum*): Sporadic throughout Mexico, Central America, and the Amazon. The transverse plane (*left callout*) shows simple and multiple pores. The banded parenchyma are characteristic for this species. The tangential section (*right callout*) shows a high number of crystals in the parenchyma cells, as well as prismatic crystals in the ray cells.

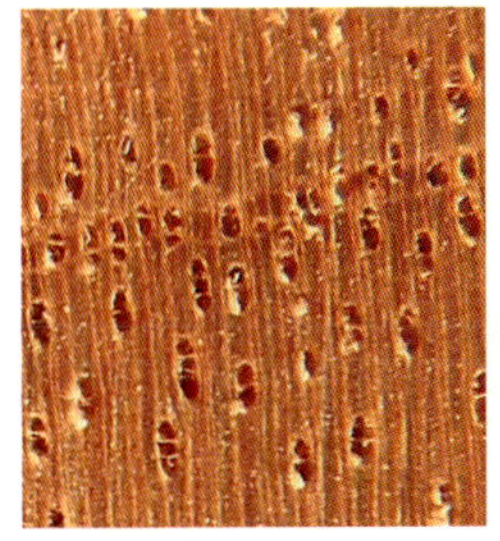

cumala (*Virola* sp.): Throughout the Amazon rainforest. Transverse plane (*left callout*) shows multiple pores, vasicentric parenchyma (with only one row of cells), and a distinct growth ring boundary. The tangential section (*right callout*) shows unicellular (rays containing upright cells only) and multicellular rays containing upright and procumbent cells, which is a distinct ID characteristic. A detail of the radial section located on the upper left of the image showcases the crystal inclusions present in the ray cells.

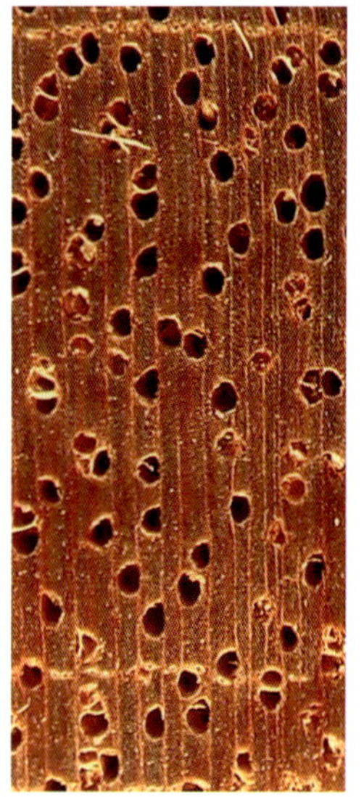

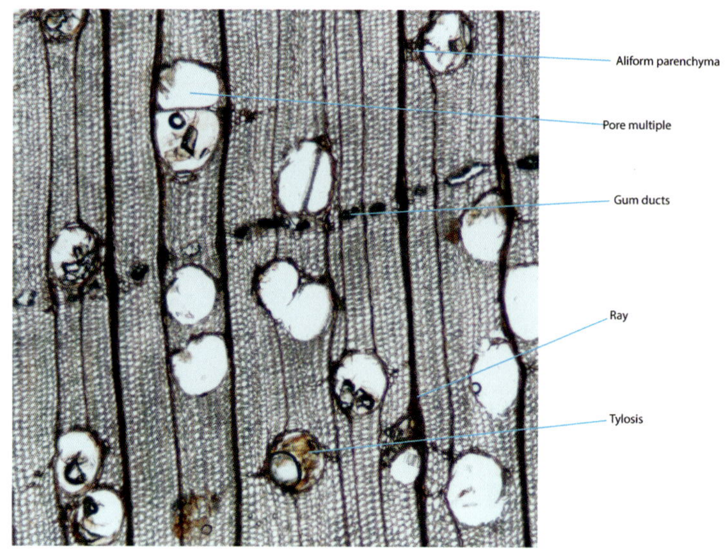


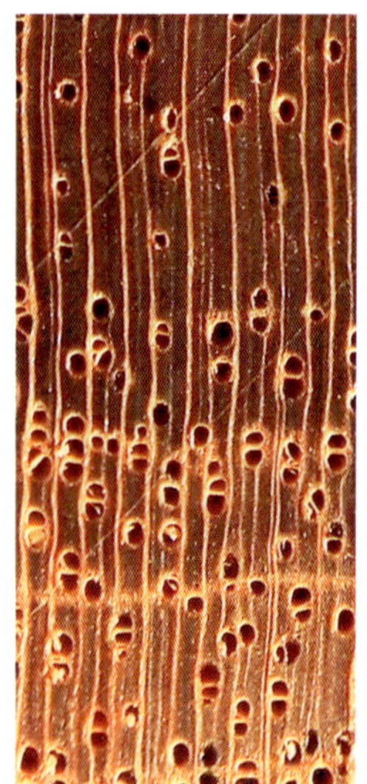

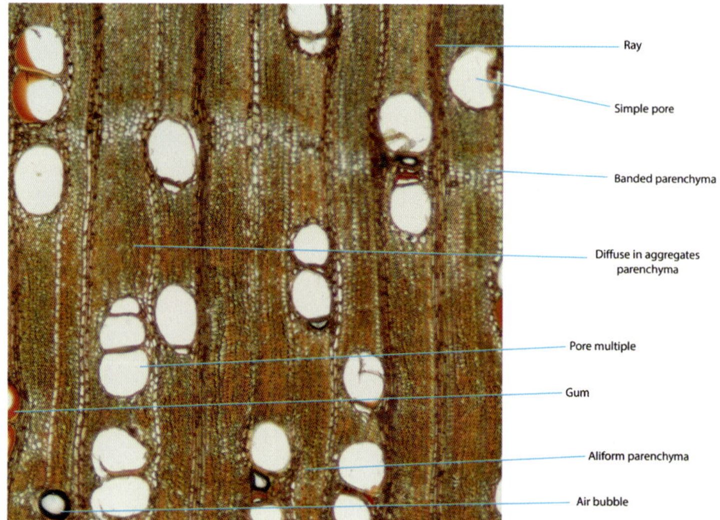


lauan / mahogany, false (*Shorea* sp.): The *Shorea* genus is wide and heavily varied. The species are sold under a number of common names, such as lauan and Philippine mahogany. Shown are two distinct species, both of which are under the *Shorea* genus. Some differences in the micrographs are the chains of gum ducts in one versus the banded parenchyma in the other. Although the gum ducts are characteristic for this genus, is possible that genetic variation causes minimal or lack of presence of this anatomical characteristic.

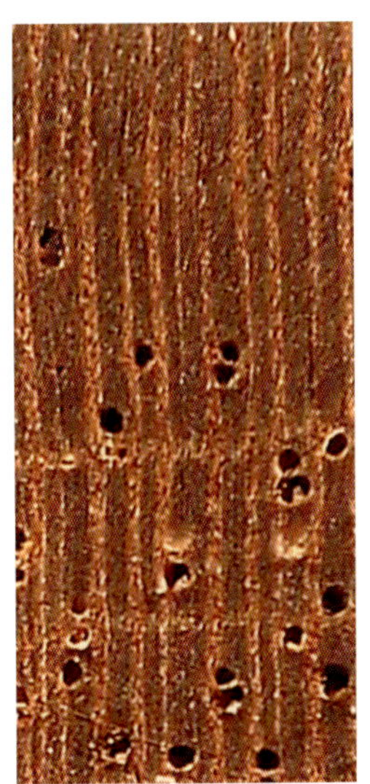

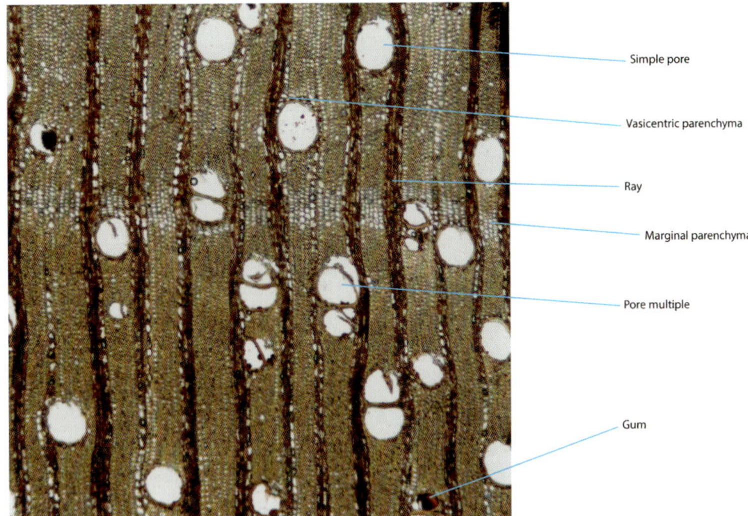


mahogany, true (*Swietenia macrophylla*): Central America and Mexico, but also found in Hawaii and throughout the world in plantations. Transverse plane shows marginal and vasicentric parenchyma. It should be noted that although there are simple and multiple pores, the last ones are predominant. This last characteristic is useful to differentiate true mahogany from Spanish cedar. Is also possible to observe red gum in the pores.

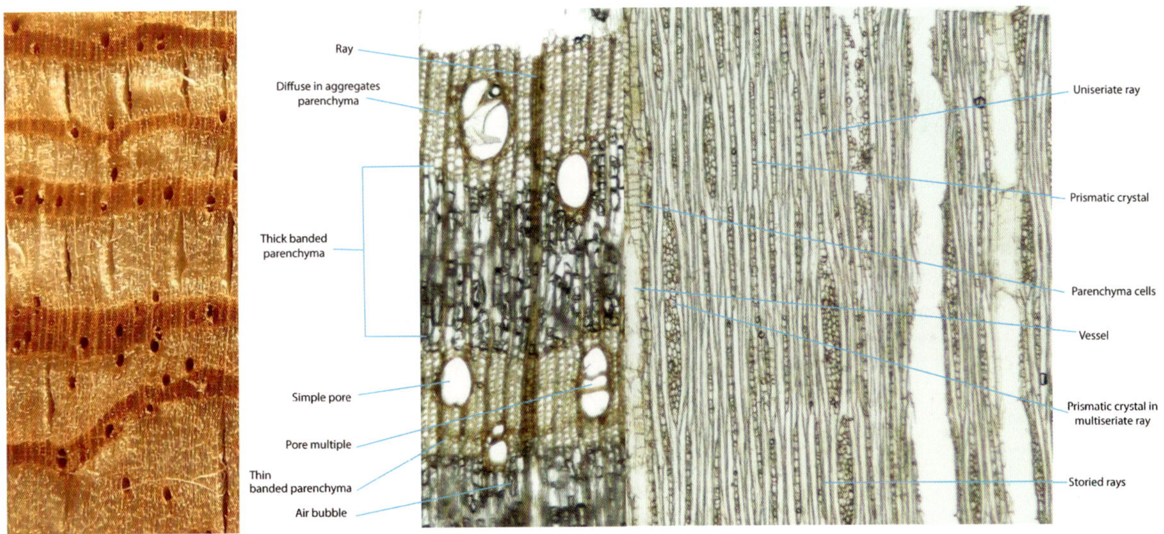


maquisapa ñaccha (*Apeiba membranacea*): North and South America. Transverse plane (*left callout*) shows thick-banded parenchyma. Other types of parenchyma also present, such as diffuse in aggregates and thin banded. The tangential section (*right callout*) shows storied rays. The rays contain prismatic crystals.

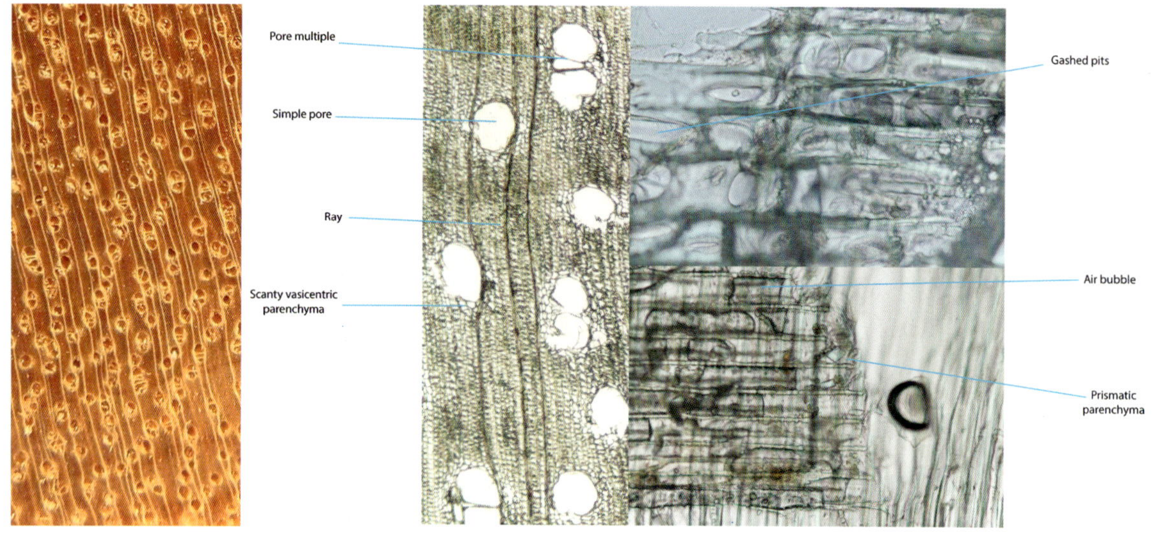


morena amarilla (*Aniba amazonica / Aniba puchury-minor*): Throughout parts of South America. Transverse plane (*left callout*) shows predominantly pore multiples, and scanty (partial) vasicentric parenchyma. On the upper right is gashed pitting between the vessel and the ray cells. On the lower left, prismatic crystals can be observed in the ray cells.

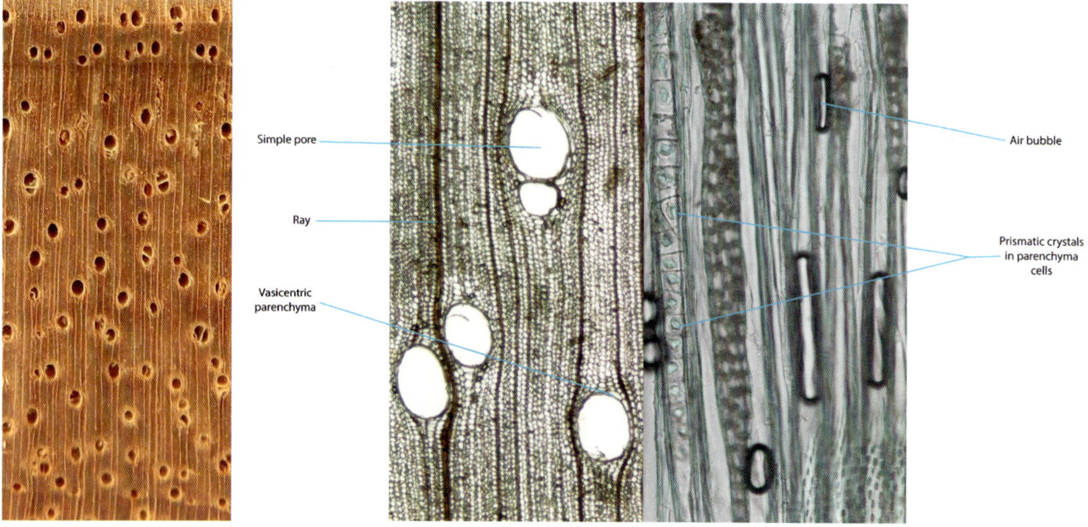


pashaco (*Macrolobium acaciaefolium*): Amazon rainforest. Transverse plane (*left callout*) shows simple pores and vasicentric parenchyma. The tangential section (*right callout*) shows prismatic crystals chambered in each parenchyma cell.

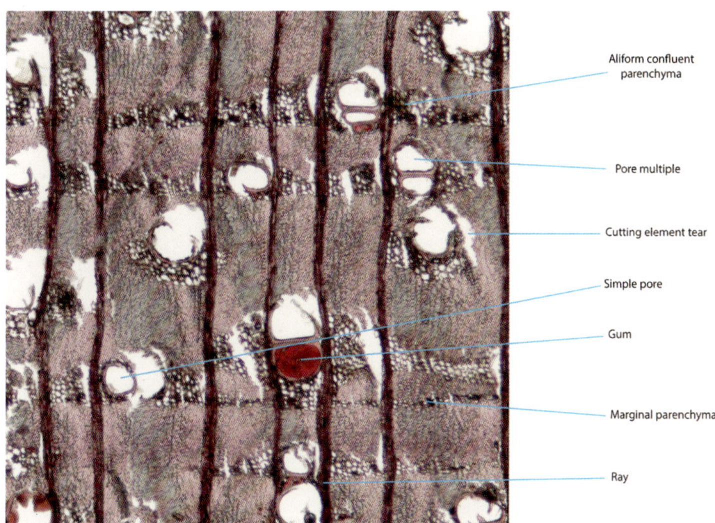


purpleheart (*Peltogyne* sp.): Rainforests of Central and South America. Transverse plane shows the characteristic purple coloration that darkens to brown/black over time. Two types of parenchyma are visible: aliform confluent and marginal. The pores are simple and multiple, and some contain red gum.

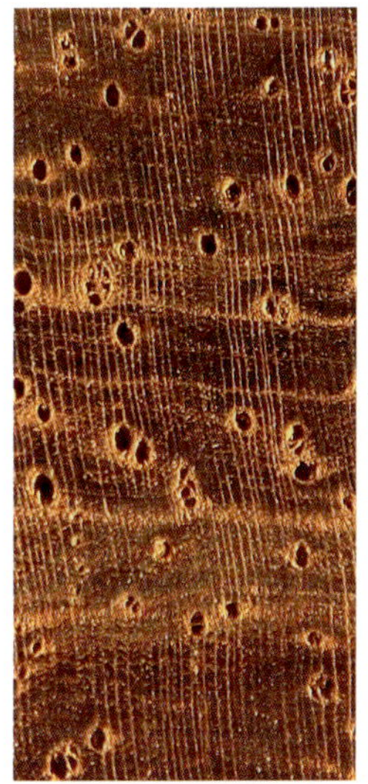

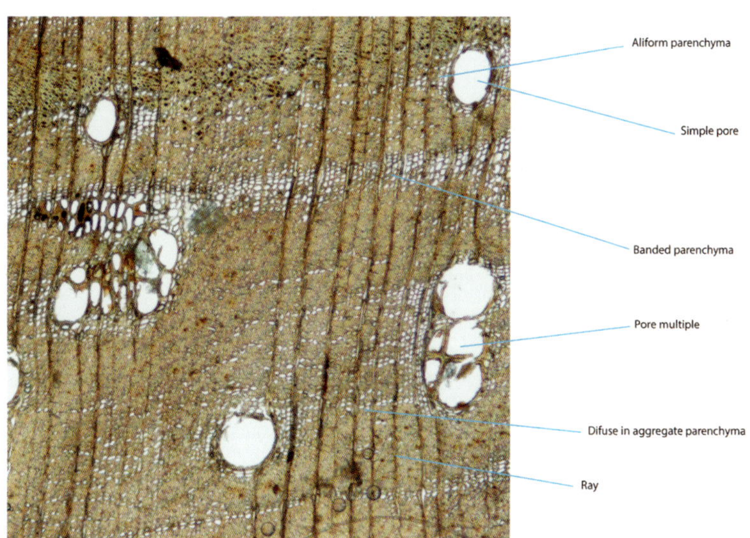


rosewood (*Dalbergia* sp.): Widely distributed across most tropical areas. The transverse plane shows aliform, banded, and diffuse-in-aggregate parenchyma, and many dark-colored extractives. Pores appear in solitary and pore multiples. Rays are primarily uniseriate.

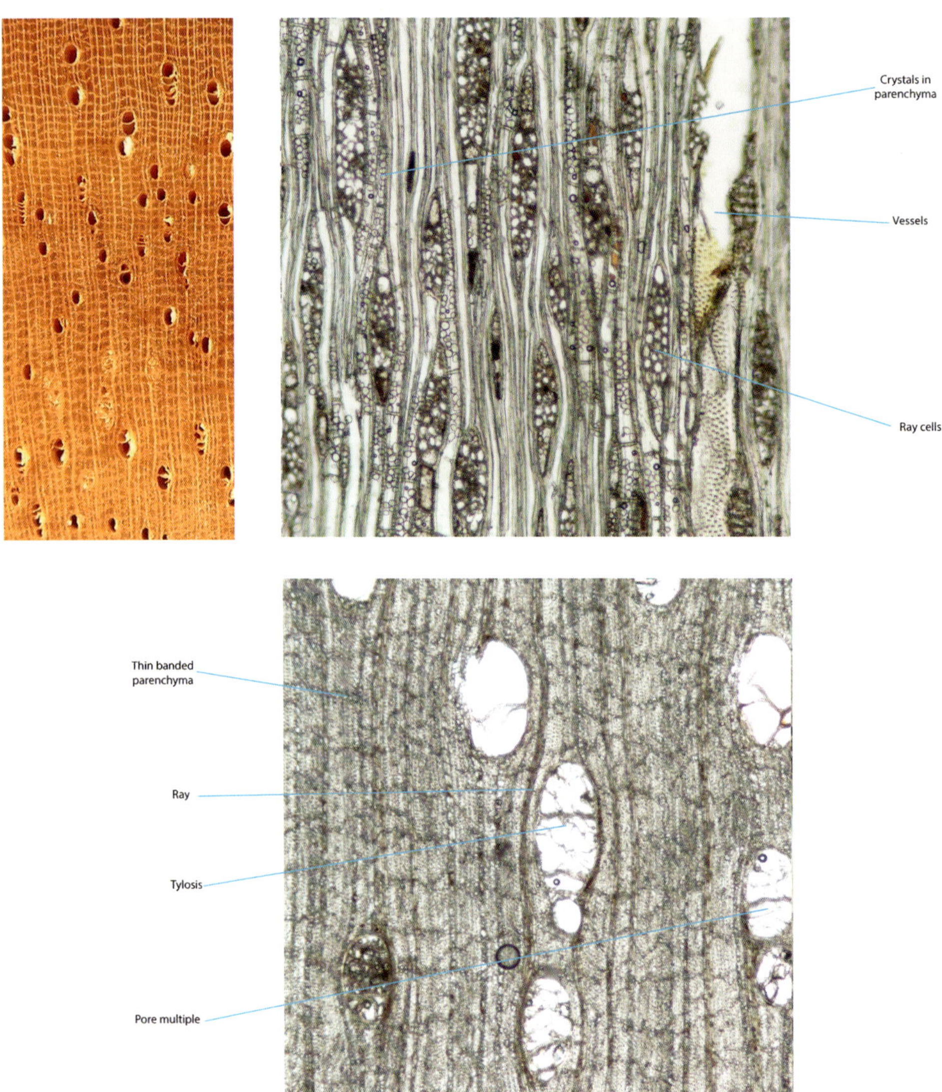


rubberwood (*Hevea brasiliensis*): South America, South and Southeast Asia. Transverse plane (*top callout*) shows tyloses in the vessels, and thin-banded parenchyma. The tangential section (*bottom callout*) shows several crystalline inclusions in the parenchyma cells.

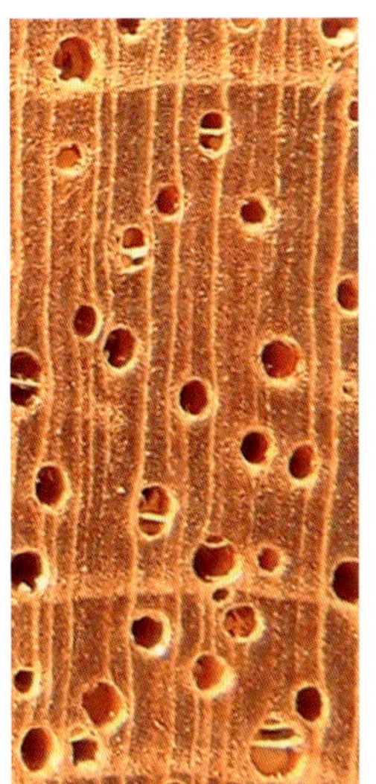

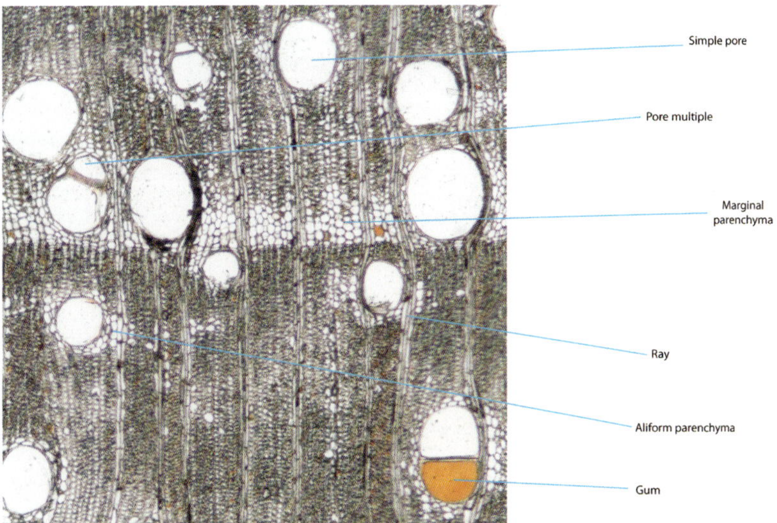


Spanish cedar (*Cedrela odorata*): In subtropical and tropical patches across Mexico and Central and South America. Transverse plane shows mostly simple pores, marginal and aliform parenchyma. It is easy to differentiate Spanish cedar from true mahogany by the pores—Spanish cedar has primarily simple pores, while true mahogany has primarily pore multiples.

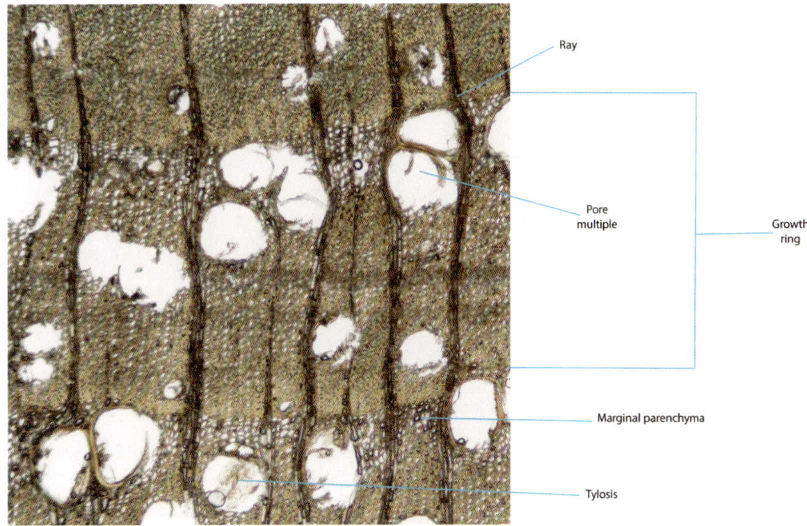


teak (*Tectona grandis*): Found throughout Asia, Africa, and the Caribbean. Transverse plane shows simple and multiple pores in a ring porous arrangement. Some of the pores show the presence of tyloses.

# Appendix IV

## NEXT STEPS IN READING

If you found this book interesting and would like to continue to learn more about wood identification and wood science, below are some next-stage readings you might consider.

## BOOKS

*Understanding Wood* (2000) by R. Bruce Hoadley is a slightly more formal introduction to wood science than this book but is also written with the crafter in mind.

*Identifying Wood* (1990) by R. Bruce Hoadley is a great introductory text on basic wood identification. It takes the "Quick and Dirty Wood ID" appendix a step further and walks the reader through some of the more specialized cells and their arrangements, giving a broad overview of hardwood, softwood, and tropical identification.

*Forest Products and Wood Science: An Introduction* by Jim Bowyer et al. (any edition is fine) takes the science one step further. Generally used as an introductory wood science textbook in universities, it offers a reasonably broad overview of wood anatomy and chemistry, as well as a look at some of its global uses.

*Spalted Wood: The History, Science, and Art of a Unique Material* (2016) by Sara Robinson, Hans Michaelsen, and Julia Robinson, covers the history and modern science of spalted wood—tracing the use of the art form from the 1300s to the modern day. It also includes a DIY guide to making your own spalted wood and using spalted wood dyes on wood products.

*Quellentexte zum Färben des Holzes 1770-1930* by Hans Michaelsen looks at the history of wood coloration in Europe, from natural to synthetic dyes and everything in between. The book is in German, but well worth it just for the images if you can't read the language.

## WEB

This book grew out of the hive-mind Facebook group Wood Education and Safety. There you can find thousands of other like-minded individuals and get real-time answers to wood science questions.

www.facebook.com/groups/woodeducation/

The main website for spalting research has lists of additional readings and events that are open to the public who wish to learn more about spalted wood and its various applications. There is also a Facebook group.

www.northernspalting.com
www.facebook.com/groups/spalting/

# Appendix V

## BASIC WOODWORKING MACHINES, TOOLS, AND THEIR APPLICATIONS

If you're just getting into woodworking, the sheer volume of tools and machines can be overwhelming. Below is a *very* simplified introduction to some of the more common woodworking machines and tools. Some brand recommendations are also provided, but note that your needs may be met with a different brand, depending on application.

**Common hand tools in a woodshop.** *Photo courtesy Brian A. Tan of Development Depot*

**bandsaw**

A bandsaw is a saw of varying sizes with a blade that is a continuous band. The band sits on two wheels: one at the top of the machine and one at the bottom. Bandsaws are very versatile and, for the hobbyist woodworker, the one piece of machinery that really matters. It is worth the money to invest in a top-quality bandsaw from the start. This machine will be with you for the rest of your life, so don't skimp on quality.

For medium to large work, Laguna bandsaws are inexpensive and very high quality. They make a menagerie of sizes and horsepower to accommodate a shop garage or a full-scale workspace. The most versatile for both spaces is www.lagunatools.com/bandsaws/1412-Bandsaw.

For small work, a small, bench-top bandsaw is great. Some of the best models have stopped being made, and there does not appear to be a community consensus on the best tabletop currently on the market. Because bandsaw quality is so highly variable, consider spending a bit more on a top-quality tabletop bandsaw instead of a cheap one, in the hope that it will last many years instead of just one.

**table saw**

A table saw is a flat, wheel-like circle with sharp edges that sticks out of a metal body. A table saw generally is used to cut wood parallel to the grain (to width), although more-advanced users can use it for cross-grain cuts for tasks such as joinery.

The saw can be dangerous for beginners if a riving knife is not used to help control kickback. There is also danger of cutting off fingers, but the Sawstop-brand table saw has addressed this issue with their machine that senses flesh and stops cutting. Sawstop is now the benchmark table saw in schools due to its safety, but be aware that the Sawstop does not prevent kickback. As such, a riving knife should still be used.

Bandsaws and scroll saws are used to cut angles and curves such as those shown on this wooden puzzle. The puzzle is maple with acrylic paint and a beeswax polish. *Photo courtesy Red Barn Toys: www.redbarn.toys*

You can buy wood planed and jointed, but it is costly. If you have the machines at home, the way to get those nice straight edges that are 90 degrees to the (flat) faces is through a process called "squaring up stock." The process is as follows: (1) cut wood to rough length on a chop saw, (2) joint one edge on the jointer (concave side toward the fence), (3) face joint one face (concave side down), (4) surface the convex side with the surface, (5) rip to width on the table saw, (6) cross-cut to exact length on a chop saw. *Photo courtesy Frank Varro, as part of a crib he built for his son*

### jointer

A jointer is a set of cylindrical-like blades set into a metal bed. The bed has a fence that is normally set 90 degrees to the bed, so that wood edges can be machined to be 90 degrees to the broad face. It is critical for the process of squaring up stock, which is where a board is taken from whatever proportions it has upon drying to four faces all 90 degrees to one another. For furniture makers, a jointer is a critical piece of equipment. Makers of small toys, as well as woodturners, may find the jointer superfluous for their shop needs. If rigged up properly and if the user accounts for drift, a bandsaw can be used in place of a jointer, although it's not as easy as one might think.

Especially for furniture applications, boards must be square and flat to fit properly. Ninety-degree angles are critical, especially for joinery. *Photo courtesy Frank Varro, as part of a crib he built for his son*

### planer

A planer is a lot like a jointer, except the wood passes through a complete housing and the blades are on top. On the bottom is a moving bed that the wood is placed on and fed through the machine. Planers—also called surfacers—are used to get the wood to a certain thickness. They are also critical in the process of squaring up stock, and planing the broad surfaces of the wood 90 degrees to the jointed edge(s). A good planer is critical in a furniture shop but has few uses in a wood-turning shop.

Live-edge vessel turning on a Delta midi lathe. Big leaf maple with white rot and black zone line-type spalting. *Turned by Letty Camire*

### lathe

A lathe is a machine unlike any other in the shop. Instead of having a blade, it simply holds the wood between two horizontal centers and spins it very quickly. It is up to you, the woodworker, to shape the wood by using various hand tools. Lathes come in a large variety of sizes and styles. For beginners, small pen lathes are usually the best bet. If you think you are really going to enjoy turning but don't want to invest a lot, there is a "midi" line of lathes that are easier on the pocketbook and still quite small and portable. For the serious woodturner there are large, "never move them once you've picked a spot" lathes.

The midi lathe market used to be fairly competitive, but in recent years a clear winner has arisen. Jet makes the now benchmark standard of midi lathe, available in three different configurations (small manual speed, small variable speed, and large variable speed). For the beginning turner or traveling turner, the small manual-speed Jet lathe cannot be beat. It is nearly indestructible and has very few components to break. It is lightweight enough to be easily picked up, it fits in most car trunks, and it can sit on a basic folding table for demos. Its short swing makes it very stable for turning woods such as spalted wood, and its compact size makes it easy to store in any shop.

Numerous spindle lathes are now on the market, and there is little difference between them. New buyers will find themselves pulled between camps, specifically the Powermatic versus Oneway fans. Your needs as a turner are more important than brand in this arena, so consider what you actually want to turn—both now and in the future—before making your purchase. Don't let yourself be swayed into a more expensive lathe just because of the brand name.

**chop saw**

A chop saw has a blade similar to a table saw, but it is housed in a movable head that can be pulled down and slid across the wood. Chop saws are meant to cut wood perpendicular to the grain, and many (often called compound miter saws) have a tilting head so you can cut wood at an angle (for something such as picture frames). Chop saws are generally inexpensive and have a number of excellent uses. Most shops use chop saws to some degree.

Squared-up stock that has been cut to length on a chop saw. *Photo courtesy Brian A. Tan of Development Depot*

**jigsaw**

This is not a common machine for bigger-scale woodworkers but is common for hobbyist woodworkers. A jigsaw has a thin sliver of a blade that is (usually) held between two centers vertically and is pulled up and down by the machine. These saws are not capable of cutting thick pieces of wood and are instead meant for thinner craft projects, such as making puzzles and sawing intricate curves. Many new toymakers invest in jigsaws but quickly find that they require more power. The correct blade on a bandsaw can accomplish most of what a jigsaw can, so unless you are doing marquetry work, consider buying a small bandsaw instead of a jigsaw.

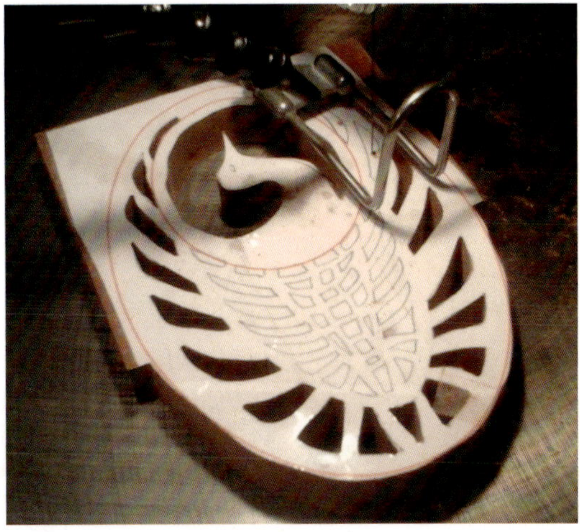

A scroll saw cuts intricate, tight curves in the interior of wood. This is accomplished due to the thinness of the blade and because the blade is a straight shaft, instead of a band such as in a bandsaw. *Image courtesy Sarah Jensen*

The perfect circular holes in this lacing toy were made with a simple drill bit in a drill press. *Image courtesy Kati Owens Gonzales: www.redbarn.toys*

### drill press

A drill press holds a drill bit in a chuck. You can then lower the head via a lever and drill a pretty straight hole into a piece of stationary wood. These machines don't have a lot of use for the hobbyist, but professional woodworkers use them a lot. Woodturners often turn drill presses into stationary sanders.

A drill press from the top down, showing the handle (*right*, with the black knob), the chuck (the large silver shaft), and the bit (a Forstner bit) in action. *Image courtesy Steve Strom*

Typical bench grinder with CBN wheels

**stationary sanders**

There are many types of stationary sanders, and oftentimes you can purchase combo units that combine several into one machine. The two most common types are belt sanders, which have the sandpaper in a continuous belt, and disc sanders, on which the sandpaper is a circle and rotates on a disc. Less frequently used stationary sanders include oscillating spindle sanders (sandpaper moves up and down on a spindle) and wide belt sanders (same deal as a normal belt sander, but on a massive scale and meant for sanding tabletops).

If you often make work that is flat on one side, an upright belt sander is handy because it can sand an entire plane at once. For portable sanding, disc sanders tend to be used more for fine grinding. Random-orbit sanders are great for sanding smaller or more delicate flatwork, whereas hand-held belt sanders are great for refinishing flooring (but very aggressive for anything else).

**bench grinders**

Bench grinders are in a small, portable housing with usually two abrasive wheels. The abrasive can vary, with the newest on the market being CBN (cubic boron nitride). Bench grinders are used primarily for tool sharpening and buffing. They come in a variety of speeds, sizes, and models, from the Tormek system to a woodturner's slow-speed grinder.

For woodturners and wood-turning tools, a slow-speed grinder is recommended. These also often come with a different type of wheel that is nicer for the high-end steel common in wood-turning tools.

## Hand Tools

A rack of C-clamps. These types of clamps have a specific (if not somewhat limited) use and are very handy when gluing boards side to side. C-clamps can be placed on flattening bars across the broad face of the glue-up, keeping the sides flush. *Photo courtesy Brian A. Tan of Development Depot*

A rack of bar clamps. These clamps tend to be used to span long distances, especially on side-to-side glue-ups. *Photo courtesy Brian A. Tan of Development Depot*

An array of hand planes. Different hand planes are for different purposes. With the right kinds, almost all squaring up can be done without large machines, but there is a huge time investment. They are most frequently used in cabinet-making and other flat to cube work, where glued-up areas must sit flush. Many instrument makers use planes for fine shaping. *Photo courtesy Brian A. Tan of Development Depot*

Handsaw in use. There are many, many types of handsaws, from Western-style push saws to Japanese pull saws. There are dovetail saws, crosscut saws (pictured here), general joinery saws, and so many more. All have their specific purposes, but generally with saws, you get what you pay for. Make sure you purchase the appropriate saw for your needs, and that you check to see which way it cuts before using it. *Photo courtesy Brian A. Tan of Development Depot*

Drill bits of all forms are a must-have for a shop. They come in every size and configuration you can think of. Instead of buying a large assortment, buy as you need them for projects. Make sure whatever you buy is for the kind of drill you have. Drill bits meant for a hand drill sometimes won't work in a drill press, and vice versa. *Image courtesy Candace Hermann: https://stayingin-courage.com*

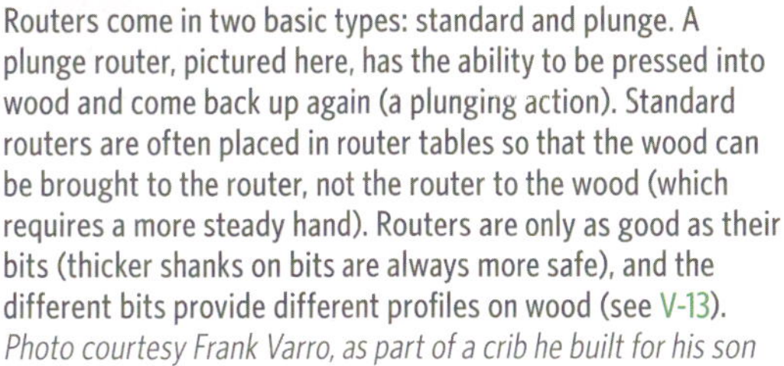

Routers come in two basic types: standard and plunge. A plunge router, pictured here, has the ability to be pressed into wood and come back up again (a plunging action). Standard routers are often placed in router tables so that the wood can be brought to the router, not the router to the wood (which requires a more steady hand). Routers are only as good as their bits (thicker shanks on bits are always more safe), and the different bits provide different profiles on wood (see V-13). *Photo courtesy Frank Varro, as part of a crib he built for his son*

Sides of a crib that have been rounded over with a round-over bit on a router. *Photo courtesy Frank Varro, as part of a crib he built for his son*

## SAFETY

Safety is of paramount importance when woodworking. Wood dust, regardless of species, is a known human carcinogen and can be very irritating even in small amounts. Even with hand tools, wood can chip and fly into the eyes, leading to irritation, scratching, and sometimes blindness. Different woodworking actions require various levels of protection, and a pictorial guide is included below to aid in selecting appropriate safety gear.

Even basic clothing can be impractical in the woodshop. Scoop-necked shirts allow wood chips to slide into clothing. For those who wear bras, the chips can get inside and scrape sensitive skin, sometimes even leading to bleeding. All adornments on the wrists and fingers should be removed, if possible.

A much-better shirt for woodworking, with a tighter neck to control introduction of wood chips. The clothing isn't so loose that it could get caught in machines. For wood turning on a lathe in particular, short sleeves should be worn. A ring is still on the model's finger and should be removed before woodworking.

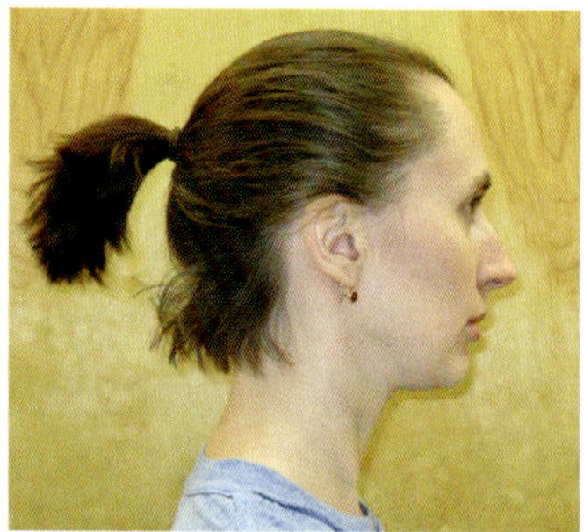

Long hair should be pulled back and secured. If your hair is longer than about shoulder length, consider wearing a hat, since strands can come loose from braids or tails and wrap into woodworking equipment. Earrings are generally fine to stay as long as they don't dangle past the mouth.

Height is critical when using machines. If a machine is too tall for the user or too short, it becomes a safety issue. Here, the model demonstrates appropriate lathe height. A wood lathe should sit so that the headstock spindle is directly in line with a 90-degree angle formed by the arm (as pictured). This lathe is too tall for the model to use easily. Note that some people prefer their lathe a bit higher or lower than the standard, and that is fine. What should be avoided is having to stand on tiptoes to use it, or bending over so much that clothing could come in contact with the spindle.

A lathe at the correct height for the model

A lathe just slightly too high for the model, but still safe to use if the user finds the height more appropriate.

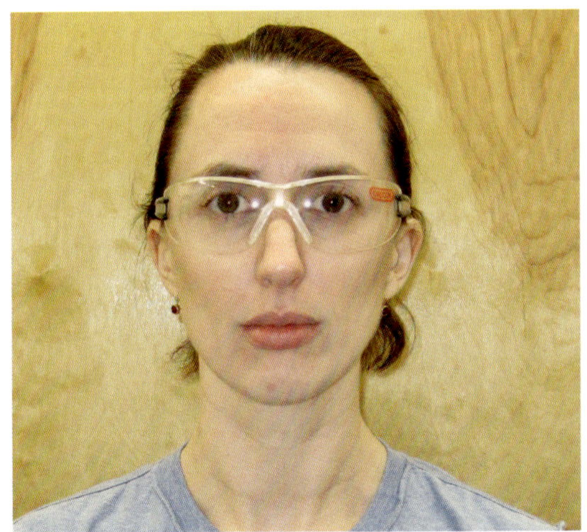

A basic pair of safety glasses is the bare minimum that should be worn in a woodshop. Upon entering a workspace, safety glasses should be worn consistently and not removed until exiting the space. For those who wear glasses, prescription safety glasses can be ordered from most optometrists.

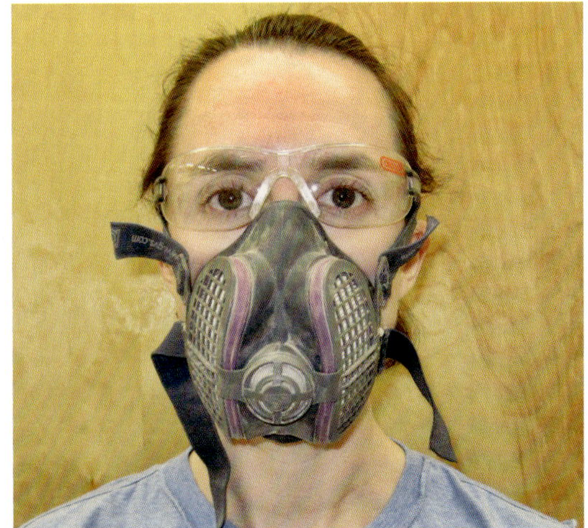

Respirators can filter particulate, vapors, or both, depending on the type of filter. Pictured here is a basic face respirator with NIOSH 95 filters on either side, The respirator modeled is small and unobtrusive, and ideal for those with smaller faces. Filters should be replaced as needed. Larger systems come with forced air and face shields, but buyers should inspect the rating of the filters before purchasing, since a respirator should have at least a NIOSH 95 rating for use in a woodshop. For use in a finishing room, filters meant for organic vapors should be used. The two types of masks are not interchangeable, but some setups do come with the same headpiece, and filters that can be easily swapped out. Respirators should also be nonnegotiable in a shop space, just like safety glasses.

Face shields are good when using wood lathes or other tasks where the wood has a chance of flying back at your face. Get a good-quality face shield with a top component that covers at least the front half of your head, as pictured here.

Addition of a hard hat to the safety gear. For turning large-scale or sculptural work, a hard hat is a good choice.

Side profile of the hard hat with face shield adapter. Safety glasses and respirator are still worn under the face shield. While the shield is impact rated, wood flying off the lathe will still press the shield into the face. Having safety glasses and a respirator underneath helps protect the face from the broad impact.

# Appendix VI

## WOOD SCIENCE EDUCATION PROGRAMS IN THE UNITED STATES

If you are interested in pursuing wood science (or a related field) at the university level, below is a list of wood science programs in the United States. Many, *many* more programs exist outside the US, but since wood science is an increasingly rare program of study in the US, and this book has a US focus, here we will focus on just US programs and the bigger Canadian programs. This list is *not* comprehensive.

Here is some general history about wood science programs in North America:

www.swst.org/meetings/AM11/pdfs/Armstrong.pdf

## UNIVERSITY LIST

Auburn University: Sustainable biomaterials and packaging. https://sfws.auburn.edu/sustainable-biomaterials-packaging/

Louisiana State University: School of Renewable Natural Resources. www.rnr.lsu.edu/

North Carolina State University: Department of Forest Biomaterials. Degrees in paper science & engineering *and* sustainable materials and technology. https://cnr.ncsu.edu/fb/

Oregon State University: Department of Wood Science & Engineering. Degree in renewable materials, with four options: advanced wood manufacturing, art & design, marketing & management, and science & engineering. The department also has a new program (master of design) coming online in 2020 for those who wish to blend wood science with wood art and design. www.forestry.oregonstate.edu/undergraduate-programs/renewable-materials

Purdue University: Forestry and Natural Resources, major in sustainable biomaterials: process & product design. https://ag.purdue.edu/fnr/Pages/ossSUBO.aspx

University of British Columbia: Department of Wood Science, BS in wood products processing. http://wood.ubc.ca

University of Idaho: Department of Forest, Rangeland & Fire Sciences, major in renewable materials. www.uidaho.edu/cnr/departments/forest-rangeland-and-fire-sciences

University of Maine: School of Forest Resources, BS in forest operations, bioproducts, and bioenergy. https://forest.umaine.edu/undergraduate-programs/

University of Massachusetts Amherst: Building and Construction Technology, Department of Environmental Conservation. MS and PhD in sustainable building systems. http://bct.eco.umass.edu/academics/graduate-studies-sustainable-building-systems/wood-engineering-specialization/

University of Toronto: Faculty of Forestry, major in forest biomaterials science. http://forestry.utoronto.ca/undergraduate-programs/

Virginia Tech: Degree in sustainable biomaterials, with the following tracks: forest products business, wood materials science, sustainable residential structures, and creating sustainable society. They also offer another degree in packaging systems and design. http://sbio.vt.edu

West Virginia University: Division of Forestry and Natural Resources, major in wood science and technology. https://forestry.wvu.edu/undergraduate/majors/wood-science-and-technology

# General Bibliography


1. Ak, N. O., D. O. Cliver, and C. W. Kaspar. "Cutting Boards of Plastic and Wood Contaminated Experimentally with Bacteria." *Journal of Food Protection* 57, no. 1 (1994): 16–22.
2. Cliver, D. O.. "Cutting Boards in *Salmonella* Cross-Contamination." *Journal of AOAC International* 89, no. 2 (2006): 538–42.
3. Dayawansa, S., K. Umeno, H. Takakura, E. Hori, E. Tabuchi, Y. Nagashima, H. Oosu, et al. "Autonomic Responses during Inhalation of Natural Fragrances of "'Cedrol' in Humans." *Autonomic Neuroscience* 108, nos. 1-2 (2003): 79–86.
4. Dixon, Matt. "$750K Sculpture Sickened FBI Workers in Miami." *Politico*, December 2, 2016. www.politico.com/states/florida/story/2016/12/the-750k-sculpture-that-hospitalized-fbi-miami-workers-107768 (last accessed December 29, 2018).
5. Flentke, George. *The Dangers of Softwood Shavings*. https://rabbit.org/care/shavings.html (last accessed December 29, 2018).
6. Khine, H., D. Weiss, N. Graber, R. S. Hoffman, N. Esteban-Cruciani, and J. R. Avner. "A Cluster of Children with Seizures Caused by Camphor Poisoning." *Pediatrics* 123, no. 5 (2009). http://pediatrics.aappublications.org/content/123/5/1269 (last accessed December 29, 2018).
7. Love, J. N., M. Sammon, and J. Smereck. "Are One or Two Dangerous? Camphor Exposure in Toddlers." *Journal of Emergency Medicine* 27, no. 1 (2004): 49–54.
8. Robinson, S. C., H. Michaelsen, and J. C. Robinson. *Spalted Wood: The History, Science, and Art of a Unique Material.* Atglen, PA: Schiffer, 2016.
9. Sano, A., H. Sei, H. Seno, Y. Morita, and H. Moritoki. "Influence of Cedar Essence on Spontaneous Activity and Sleep of Rats and Human Daytime Nap." *Psychiatry and Clinical Neurosciences* 52, no. 2 (1998): 133–35.
10. Segelman, A. B., F. P. Segelman, J. Karliner, and D. Sofia. "Sassafras and Herb Tea: Potential Health Hazards." *Journal of the American Medical Association* 236, no. 5 (1976): 477.
11. True, R. G., and J. E. Lowe. "Induced Juglone Toxicosis in Ponies and Horses." *American Journal of Veterinary Research* 41, no. 6 (1980): 944–45.

# Appendix I References


Bleumink, E., J. C. Mitchell, and J. P. Nater. "Allergic Contact Dermatitis from Cedar Wood (*Thuja plicata*)." *British Journal of Dermatology* 88, no. 5 (1973): 499-504.

Bonny, M., O. Aerts, J. Lambert, and H. Lapeere. "Occupational Contact Allergy Caused by Pao Ferro (Santos Rosewood): A Report of Two Cases." *Contact Dermatitis* 68, no. 2 (2013): 126-28.

Bush, R. K., J. W. Yunginger, and C. E. Reed. "Asthma Due to African Zebrawood (Microberlinia) Dust." *American Review of Respiratory Disease* 117, no. 3 (1978): 601-603.

Carlos de Sá Silva, Roberto, Julio Cesar Raposo de Almeida, Ana Aparecida da Silva Almeida, Gokithi Akisue, Matheus Diniz Gonçalves Coelho, and José Roberto Pereira. "Evaluation of the Toxicity of Araribá (*Centrolobium tomentosum*) Using Brine Shrimp Test." *Ambiente e Agua: An Interdisciplinary Journal of Applied Sciences* 8, no. 4 (2013): 158-67.

Chan-Yeung, M., and R. Abbound. "Occupational Asthma Due to California Redwood (*Sequoia sempervirens*) Dusts." *American Review of Respiratory Disease* 114, no. 5 (1976): 1027-31.

Chen, Weiyang, Ilze Vermaak, and Alvaro Viljoen. "Camphor—a Fumigant during the Black Death and a Coveted Fragrant Wood in Ancient Egypt and Babylon—a Review." *Molecules* 18, no. 5 (2013): 5434-5454.

Ferreira, O., M. João Cruz, A. Mota, A. Paula Cunha, and F. Azezedo. "Erythema Multiforme-Like Lesions Revealing Allergic Contact Dermatitis to Exotic Woods." *Cutaneous and Ocular Toxicology* 31, no. 1 (2011): 61-63.

Ferrer, A., F. Marañón, M. Casanovas, and E. Fernández-Caldas. "Asthma from Inhalation of *Triplochiton scleroxylon* (Samba) Wood Dust." *Journal of Investigational Allergology & Clinical Immunology* 11, no. 3 (2001): 199-203.

Gómez-Muga, S., J. A. Ratón-Nieto, and I. Ocerin. "An Unusual Case of Contact Dermatitis Caused by Wood Bracelets." *Contact Dermatitis* 61, no. 6 (2009): 351-52.

Hausen, Björn M. *Woods Injurious to Human Health: A Manual*. Berlin and New York: Walter de Gruyter, 1981.

Hausen, B. M., G. Bruhn, and W. A. Koenig. "New Hydroxyisoflavans as Contact Sensitizers in Cocus Wood *Byra ebenus* DC (Fabaceae)." *Contact Dermatitis* 25, no. 3 (1991): 149-55.

Huilaja, L., M. E. Kubin and R. Riekki. "Contact Allergy to Finished Woods in Furniture and Finishings: A Small Allergic Contact Dermatitis Epidemic to Western Red Cedar in Sauna Interior Decoration." *Journal of the European Academy of Dermatology and Venereology* 30, no. 1 (2016): 57-59.

Soderquist, Charles J. "Juglone and Allelopathy." *Journal of Chemical Education* 50, no. 11 (1973): 782-83.

Woods, Brian, and C. D. Calnan. "Toxic Woods." *British Journal of Dermatology* 94, no. 13 (1976): 1-97.

"Woods [MAK Value Documentation, 1999]." In *The MAK-Collection for Occupational Health and Safety: Annual Thresholds and Classifications for the Workplace*. Edited by Helmut Greim, 284-89. Weinheim, Germany: Wiley-VCH Verlag, 2012.

"Woods: *Terminalia ivorensis* A. Chev. [MAK Value Documentation 2002]." In *The MAK-Collection for Occupational Health and Safety*. Edited by Helmut Greim, 259-62. Weinheim, Germany: Wiley-VCH Verlag, 2012.

# INDEX TO WOOD SPECIES DISCUSSED IN THIS BOOK


A

*Abies amabilis*
*Abies procera*
acacia
*Acacia* sp.
*Acer macrophyllum*
*Acer saccharum*
alder, black
alder, European
*Alnus glutinosa*
*Aniba amazonica*
*Aniba puchury-minor*
*Apeiba membranacea*
*Arbutus menziesii*
ash, black
ash, white
aspen
*Astronium fraxinifolium*

B

balsa
beech, American
beech, European
*Betula papyrifera*
birch
black walnut
bloodwood
bolaina
breadnut
*Brosimum alicastrum*
*Brosimum rubescens*

C

cachimbo
*Calocedrus decurrens*
camphor
canary wood
*Cariniana domestica*
*Carya ovate*
*Carya* spp.
cedar, eastern white
cedar, incense
cedar, western red
*Cedrela odorata*
*Centrolobium tomentosum*
*Chamaecyparis lawsoniana*
chinkapin
*Chrysolepis chrysophylla*
*Cinnamomum camphor*
congona
*Cornus florida*
cumala

D

*Dalbergia latifolia*
*Dalbergia nigra*
*Dalbergia sissoo*
*Dalbergia* sp.
*Diosyros ebenum*
dogwood
Douglas fir

E

eastern red cedar
eastern spruce
eastern white pine
ebony
elm, American
elm, red

F

*Fagus grandifolia*
*Fagus sylvatica*
*Fraxinus americana*
*Fraxinus nigra*

G

*Guazuma crinita*

H

hemlock, mountain
*Hevea brasiliensis*
hickory

J

*Juglans nigra*
juniper
*Juniperus* sp.
*Juniperus virginiana*

K

*Khaya anthotheca*

L

larch, western
*Larix occidentalis*
lauan
limba
*Lirodendron tulipifera*
lodge pole pine
long leaf pine

M

*Machaerium* spp.
*Macrolobium acaciaefolium*
madrone
mahogany, African
mahogany, Cuban
mahogany, false
mahogany, Philippine
mahogany, true
maple, big leaf
maple, sugar
maquisapa ñaccha
*Microberlinia brazzavillensis*
morena amarilla
myrtle

N

noble fir

O
oak, red
oak, white
*Ochroma pyramidale*
*Olea europaea*
olive

P
Pacific silver fir
Pacific yew
pao ferro
pashaco
*Peltogyne densiflora*
*Peltogyne* sp.
*Picea rubens*
pine, radiata
pine, Scots
*Pinus contorta*
*Pinus palustris*
*Pinus radiata*
*Pinus strobus*
*Pinus sylvestris*
poplar
*Populus* sp.
Port Orford cedar
*Pseudotsuga menziesii*
purpleheart

Q
*Quercus alba*
*Quercus rubra*

R
redwood
rosewood
rosewood, Brazilian
rosewood, East Indian
rosewood, santos
rubberwood

S
samba
sassafras
*Sassafras albidum*
*Sequoia sempervirens*
*Shorea* sp.
sissoo
Spanish cedar
*Swietenia macrophylla*
*Swietenia mahagoni*

T
*Taxus brevifolia*
teak
*Tectona grandis*
*Terminalia superba*
*Thuja plicata*
tigerwood
*Triplochiton scleroxylon*
*Tsuga mertensiana*
tulip tree

U
*Ulmus americana*
*Ulmus rubra*
*Umbellularia californica*

V
*Virola* sp.

W
walnut

Z
zebrawood

# Glossary

bamboo: A "monocot," meaning the seed contains only one embryonic leaf. Most monocots also do not have continuous woody stems, such as dicots (trees are dicots). Bamboo, therefore, is not technically a tree, and likewise its "wood" does not behave the same as wood from a tree either.

cellulose: One of the three primary building blocks of wood. Cellulose is a long chain polymer of glucose.

cross-field: Within the confines of wood anatomy, the cross-field is the tic-tac-toe-like board that is visible under a microscope where the rays cross the longitudinal cells. In softwoods, the pits that are formed due to the rays crossing the longitudinal tracheids make unique shapes of half-bordered pits that are useful in identification.

earlywood: The springwood, or the first part of a given annual ring

gum: A type of extractive found in many hardwoods, especially in the tropics

hardwood: A deciduous tree

hemicellulose: One of the three main building blocks of wood. Hemicellulose is a branched sugar polymer.

latewood: The "summerwood," or the later sections of growth in a growth ring. The latewood cells generally have thicker cell walls than the earlywood.

lignin: One of the three main building blocks of wood. Lignin is a 3-D polymer that forms a great deal of the rigidity in wood.

parenchyma: The tree's storage cells

pit: An opening in a cell that allows for transport from side to side (into other cells)

pore: An "up and down" conductive cell in hardwoods. Also known as a vessel or vessel element.

radial plane: The plane of wood where the *sides* of the rays are visible

ray: The cells in trees that run in and out, instead of up and down like most wood cells

resin canal: An opening in a conifer, lined with epithelial cells that secrete resin (or "pitch")

softwood: A coniferous tree

tangential plane: The plane of wood in which the *ends* of the rays are visible

tracheid: A type of cell that occurs both in hardwoods and softwoods that lends both support and some conduction to the tree

transverse plane: The plane of wood in which the growth rings are visible in their entirety

tropical wood: Wood that comes from trees grown in tropical climates. Tropical wood often does not have growth rings, due to lack of seasonal cycling, or may have rings that correspond to wet and dry seasons. Tropical woods often have more extractives than temperate woods and more-specialized extractives.

vessel / vessel element: The primary conductive cell of hardwoods. Also known as "pores."

wood: Refers to the material produced by "dicots," where the seed has two embryonic leaves. Dicots often have continuous woody stems, in contrast to monocots such as bamboo and palm.

## ABOUT THE AUTHOR

**Dr. Seri C. Robinson** is a professor of wood anatomy at Oregon State University and an avid woodturner. When not begging people to stop oiling their wooden cutting boards, Robinson can be found teaching wood turning to eager undergraduates, roller skating, or extolling the beauty of spalted wood. Robinson lives with their partner and child in the Pacific Northwest and maintains a fridge filled with fungal cultures and, occasionally, food.